Using Information Technology in Mathematics Education

Using Information Technology in Mathematics Education has been co-published simultaneously as *Computers in the Schools*, Volume 17, Numbers 1/2 2001.

The *Computers in the Schools* Monographic "Separates"

Below is a list of "separates," which in serials librarianship means a special issue simultaneously published as a special journal issue or double-issue *and* as a "separate" hardbound monograph. (This is a format which we also call a "DocuSerial.")

"Separates" are published because specialized libraries or professionals may wish to purchase a specific thematic issue by itself in a format which can be separately cataloged and shelved, as opposed to purchasing the journal on an on-going basis. Faculty members may also more easily consider a "separate" for classroom adoption.

"Separates" are carefully classified separately with the major book jobbers so that the journal tie-in can be noted on new book order slips to avoid duplicate purchasing.

You may wish to visit Haworth's website at . . .

http://www.HaworthPress.com

. . . to search our online catalog for complete tables of contents of these separates and related publications.

You may also call 1-800-HAWORTH (outside US/Canada: 607-722-5857), or Fax 1-800-895-0582 (outside US/Canada: 607-771-0012), or e-mail at:

getinfo@haworthpressinc.com

Using Information Technology in Mathematics Education, edited by D. James Tooke, PhD, and Norma Henderson, MS (Vol. 17, No. 1/2 2001). *"Provides thought-provoking material on several aspects and levels of mathematics education. The ideas presented will provide food for thought for the reader, suggest new methods for the classroom, and give new ideas for further research." (Charles E. Lamb, EdD, Professor, Mathematics Education, Department of Teaching, Learning, and Culture, College of Education, Texas A&M University, College Station).*

Integration of Technology into the Classroom: Case Studies, edited by D. LaMont Johnson, PhD, Cleborne D. Maddux, PhD, and Leping Liu, PhD (Vol. 16, No. 2/3/4, 2000). *Use these fascinating case studies to understand why bringing information technology into your classroom can make you a more effective teacher, and how to go about it!*

Information Technology in Educational Research and Statistics, edited by Leping Liu, PhD, D. LaMont Johnson, PhD, and Cleborne D. Maddux, PhD (Vol. 15, No. 3/4, and Vol. 16, No. 1, 1999). *This important book focuses on creating new ideas for using educational technologies such as the Internet, the World Wide Web and various software packages to further research and statistics. You will explore on-going debates relating to the theory of research, research methodology, and successful practices.* Information Technology in Educational Research and Statistics *also covers the debate on what statistical procedures are appropriate for what kinds of research designs.*

Educational Computing in the Schools: Technology, Communication, and Literacy, edited by Jay Blanchard, PhD (Vol. 15, No. 1, 1999). *Examines critical issues of technology, teaching, and learning in three areas: access, communication, and literacy. You will discover new ideas and practices for gaining access to and using technology in education from preschool through higher education.*

Logo: A Retrospective, edited by Cleborne D. Maddux, PhD, and D. Lamont Johnson, PhD (Vol. 14, No. 1/2, 1997). *"This book–honest and optimistic–is a must for those interested in any aspect of Logo: its history, the effects of its use, or its general role in education." (Dorothy M. Fitch, Logo consultant, writer, and editor, Derry, New Hampshire)*

Using Technology in the Classroom, edited by D. LaMont Johnson, PhD, Cleborne D. Maddux, PhD, and Leping Liu, MS (Vol. 13, No. 1/2, 1997). *"A guide to teaching with technology that* *emphasizes the advantages of transiting from teacher-directed learning to learner-centered learning–a shift that can draw in even 'at-risk' kids." (Book News, Inc.)*

Multimedia and Megachange: New Roles for Educational Computing, edited by W. Michael Reed, PhD, John K. Burton, PhD, and Min Liu, EdD (Vol. 10, No. 1/2/3/4, 1995). *"Describes and analyzes issues and trends that might set research and development agenda for educators in the near future." (Sci Tech Book News)*

Language Minority Students and Computers, edited by Christian J. Faltis, PhD, and Robert A. DeVillar, PhD (Vol. 7, No. 1/2, 1990). *"Professionals in the field of language minority education, including ESL and bilingual education, will cheer this collection of articles written by highly respected, research-writers, along with computer technologists, and classroom practitioners." (Journal of Computing in Teacher Education)*

Logo: Methods and Curriculum for Teachers, by Cleborne D. Maddux, PhD, and D. LaMont Johnson, PhD (Supp #3, 1989). *"An excellent introduction to this programming language for children." (Rena B. Lewis, Professor, College of Education, San Diego State University)*

Assessing the Impact of Computer-Based Instruction: A Review of Recent Research, by M. D. Roblyer, PhD, W. H. Castine, PhD, and F. J. King, PhD (Vol. 5, No. 3/4, 1988). *"A comprehensive and up-to-date review of the effects of computer applications on student achievement and attitudes." (Measurements & Control)*

Educational Computing and Problem Solving, edited by W. Michael Reed, PhD, and John K. Burton, PhD (Vol. 4, No. 3/4, 1988). *Here is everything that educators will need to know to use computers to improve higher level skills such as problem solving and critical thinking.*

The Computer in Reading and Language Arts, edited by Jay S. Blanchard, PhD, and George E. Mason, PhD (Vol. 4, No. 1, 1987). *"All of the [chapters] in this collection are useful, guiding the teacher unfamiliar with classroom computer use through a large number of available software options and classroom strategies." (Educational Technology)*

Computers in the Special Education Classroom, edited by D. LaMont Johnson, PhD, Cleborne D. Maddux, PhD, and Ann Candler, PhD (Vol. 3, No. 3/4, 1987). *"A good introduction to the use of computers in special education. . . . Excellent for those who need to become familiar with computer usage with special population students because they are contemplating it or because they have actually just begun to do it." (Science Books and Films)*

You Can Do It/Together, by Kathleen A. Smith, PhD, Cleborne D. Maddux, PhD, and D. LaMont Johnson, PhD (Supp #2, 1986). *A self-instructional textbook with an emphasis on the partnership system of learning that introduces the reader to four critical areas of computer technology.*

Computers and Teacher Training: A Practical Guide, by Dennis M. Adams, PhD (Supp #1, 1986). *"A very fine . . . introduction to computer applications in education." (International Reading Association)*

The Computer as an Educational Tool, edited by Henry F. Olds, Jr. (Vol. 3, No. 1, 1986). *"The category of tool uses for computers holds the greatest promise for learning, and this . . . book, compiled from the experiences of a good mix of practitioners and theorists, explains how and why." (Jack Turner, Technology Coordinator, Eugene School District 4-J, Oregon)*

Logo in the Schools, edited by Cleborne D. Maddux, PhD (Vol. 2, No. 2/3, 1985). *"An excellent blend of enthusiasm for the language of Logo mixed with empirical analysis of the language's effectiveness as a means of promoting educational goals. A much-needed book!" (Rena Lewis, PhD, Professor, College of Education, San Diego State University)*

Humanistic Perspectives on Computers in the Schools, edited by Steven Harlow, PhD (Vol. 1, No. 4, 1985). *"A wide spectrum of information." (Infochange)*

Using Information Technology in Mathematics Education

D. James Tooke
Norma Henderson
Editors

Using Information Technology in Mathematics Education has been co-published simultaneously as *Computers in the Schools*, Volume 17, Numbers 1/2 2001.

LONDON AND NEW YORK

Using Information Technology in Mathematics Education has been co-published simultaneously as *Computers in the Schools*, Volume 17, Numbers 1/2 2001.

First published 2001 by The Haworth Press, Inc.

2 Park Square, Milton Park, Abingdon, Oxfordshire OX14 4RN
605 Third Avenue, New York, NY 10017

Routledge is an imprint of the Taylor & Francis Group, an informa business

First issued in hardback 2020

Cover design by Thomas J. Mayshock Jr.

Library of Congress Cataloging-in-Publication Data

Using information technology in mathematics education / D. James Tooke, Norma Henderson, editors
    p. cm.
    "Using information technology in mathematics education has been co-published simultaneously as Computers in the schools, volumes 17, numbers 1/2 2001."
    Includes bibliography references and index
    ISBN 0-7890-1375-4 (alk. paper) – ISBN 0-7890-1376-2 (alk. paper)
    1. Mathematics–Computer-assisted instruction. I. Tooke, D. James. II. Henderson, Norma

    QA20.C65 U78 2001
    710'.71–dc21   2001024286

ISBN 978-0-7890-1375-0 (hbk)
ISBN 978-0-7890-1376-7 (pbk)

Indexing, Abstracting & Website/Internet Coverage

This section provides you with a list of major indexing & abstracting services. That is to say, each service began covering this periodical during the year noted in the right column. Most Websites which are listed below have indicated that they will either post, disseminate, compile, archive, cite or alert their own Website users with research-based content from this work. (This list is as current as the copyright date of this publication.)

Abstracting, Website/Indexing Coverage ...... Year When Coverage Began

- *Academic Abstracts/CD-ROM* **1994**
- *ACM Guide to Computer Literature* **1991**
- *BUBL Information Service, an Internet-based Information Service for the UK higher education community, <URL:http://bubl.ac.uk/>* **1994**
- *Child Development Abstracts & Bibliography* **2000**
- *CNPIEC Reference Guide: Chinese National Directory of Foreign Periodicals* **1995**
- *Computer Literature Index* **1993**
- *Computing Reviews* **1992**
- *Current Index to Journals in Education* **1991**
- *Education Digest* **1991**
- *Education Index* **1999**
- *Education Process Improvement Ctr, Inc. (EPICENTER) <www.epicent.com>* **2000**
- *Educational Administration Abstracts (EAA)* **1991**

(continued)

*Special Bibliographic Notes related to special journal issues
(separates) and indexing/abstracting:*

- indexing/abstracting services in this list will also cover material in any "separate" that is co-published simultaneously with Haworth's special thematic journal issue or DocuSerial. Indexing/abstracting usually covers material at the article/chapter level.
- monographic co-editions are intended for either non-subscribers or libraries which intend to purchase a second copy for their circulating collections.
- monographic co-editions are reported to all jobbers/wholesalers/approval plans. The source journal is listed as the "series" to assist the prevention of duplicate purchasing in the same manner utilized for books-in-series.
- to facilitate user/access services all indexing/abstracting services are encouraged to utilize the co-indexing entry note indicated at the bottom of the first page of each article/chapter/contribution.
- this is intended to assist a library user of any reference tool (whether print, electronic, online, or CD-ROM) to locate the monographic version if the library has purchased this version but not a subscription to the source journal.
- individual articles/chapters in any Haworth publication are also available through the Haworth Document Delivery Service (HDDS).

Using Information Technology in Mathematics Education

Contents

ABOUT THE EDITORS

D. James Tooke, PhD, is Professor of Mathematics Education at Eastern Oregon University, where he teaches both mathematics and mathematics education. He received his PhD from Texas A&M University and his BS and MA in Mathematics from Sam Houston State University, and has taught at the University of Houston and the University of Nevada. His research interests are examining the impact of mathematics in the academic preparation of teachers, and searching methods of reducing mathematics anxiety, particularly in k-6 teachers.

Norma Henderson, MS, is Doctoral Candidate in Counseling and Educational Psychology with an emphasis in Educational Technology at the University of Nevada, Reno. She received her BS in Civil Engineering and an MS in Information Technology in Education from the University of Nevada, Reno. She teaches courses on the application of technology in education. Her research interests focus on integration of technology in classroom teaching/learning; program evaluation and assessment, distance education, Web page development, and statistical modeling.

INTRODUCTION

D. James Tooke

Mathematics, the Computer, and the Impact on Mathematics Education

SUMMARY. The connection between mathematics and the computer is obvious. Elementary notions of mathematics gave rise to the computer; advanced notions gave it a more powerful state. But as the computer advanced, it expanded mathematics, allowing the creation of further branches of mathematics. Then the computer affected mathematics education. It changed the mathematics curriculum, the teaching of mathematics, and even the way mathematics was learned. Recently, it has affected the training of teachers, both pre-service and in-service. This publication includes articles that address each of these changes. *[Article copies available for a fee from The Haworth Document Delivery Service: 1-800-342-9678. E-mail address: <getinfo@haworthpressinc.com> Website: <http://www.HaworthPress.com> © 2001 by The Haworth Press, Inc. All rights reserved.]*

KEYWORDS. Mathematics, computer, mathematics education, technology

D. JAMES TOOKE is Professor, School of Education and Business, Eastern Oregon University, 252 Zabel Hall, One University Boulevard, La Grande, OR 97850 (E-mail: jtooke@eou.edu).

[Haworth co-indexing entry note]: "Introduction: Mathematics, the Computer, and the Impact on Mathematics Education." Tooke, D. James. Co-published simultaneously in *Computers in the Schools* (The Haworth Press, Inc.) Vol. 17, No. 1/2, 2001, pp. 1-7; and: *Using Information Technology in Mathematics Education* (ed: D. James Tooke and Norma Henderson) The Haworth Press, Inc., 2001, pp. 1-7. Single or multiple copies of this article are available for a fee from The Haworth Document Delivery Service [1-800-342-9678, 9:00 a.m. - 5:00 p.m. (EST). E-mail address: getinfo@haworthpressinc.com].

MATHEMATICS AND THE COMPUTER

There is a trivially obvious connection between mathematics and the computer. It is, in fact, a symbiotic relationship. Without mathematics, the computer would not even exist. However, the existence and development of the computer have enhanced mathematics, allowing us to go beyond the mathematics of imagination and paper only.

The most basic of mathematical ideas, place value, made possible the creation of the very elementary computer, the abacus. The same concept provided binary mathematics, the notion of base two mathematics, which produced the simplest early computers. Base two, or binary mathematics, uses only 0 and 1 to represent numbers. The computer could read 0 as "off," and 1 as "on," and thus a series of switches made a machine represent any rational number. Later the far more sophisticated notions of graph theory and topology (that branch of mathematics which consists of concepts–such as connectedness and accessibility–which allow the development of network topology) helped the development expand into today's high-speed computers, making mathematics involved in the birth and maturation of the computer.

As the computer advanced, it changed mathematics. The computer has made the study of many branches of mathematics feasible: fields such as numerical analysis, approximation theory, operations research, and advanced statistics. Fractal geometry, though conceived earlier, had no reality until the advent of high-speed computers. This branch of mathematics has taken applications of mathematics and the computer into new areas, such as the movie industry (e.g., Disney's *Black Hole* and the second *Star Trek* movie).

But it is the usage of the computer that has impacted mathematics, of course, and that requires a consideration of the field, or discipline, of computer science. When accepting the Turing Award for excellence in computer science, Juris Hartmanis (1994) said of his field, "It is the engineering of mathematics" (p. 41). While early definitions (Knuth, 1994) said that computer science is simply the science of computers, more recently it has been said that computer science is actually the science of algorithms. Because of the speed and precision provided by the computer, the richness of algorithmic studies was not fully realized until computers were available. However, as algorithms existed before computers (Knuth, 1994), there is an argument that mathematics and computer science actually came together with the advent of the computer!

In summary, mathematics gave birth to computer science, but together they have both developed significantly. All of this has certainly had an impact on many areas of mathematics education, including the mathematics curriculum, mathematics instruction, and mathematics learning.

Mathematics Curriculum

The computer impacted the mathematics curriculum long ago when it caused the infusion of a new course called "computer math." In the 1980s this writer had the opportunity to learn and teach the BASIC programming language to high school students. It was found to be crude, by today's standards, but certainly more user-friendly than the programming languages, such as Fortran and COBOL, studied in college in the 1960s.

Apparently there is a sense in the educational community that either the curriculum should catch up with society's use of technology or perhaps suffer repercussions later when parents would eventually demand it. This is supported by a 1997 position paper which states that only by recognizing and understanding past technological achievement can educators recognize and understand the influence today's technology will have on schools (Sarmiento, 1997). This is echoed today in the literature (Ediger, 1998) and might explain, to a degree, the two-decade gap between computer science being offered at the university and in the secondary school. Knowledge that mathematicians have held for 40 years has only been in the hands of educators for half that time (Kaput, 1992).

The National Council of Teachers of Mathematics formally declared its stance on computers in the curriculum in 1987 (NCTM *News Bulletin*, 1987), echoing a similar call in 1985 by the International Commission on Mathematics Instruction (Cornu & Ralston, 1992).

Among other mathematical experiences of students affected, the curricular changes brought about by the computer were the way students could be taught to gather data and accept or reject hypotheses. In summary, it caused a restructuring of the mathematics curriculum (Cusco & Goldenberg, 1996), and the same can be said about the impact of the computer's little sibling, the hand-held calculator (Waits & Demana, 1998). The estimate is that more than a quarter of the mathematics taught before the arrival of the scientific calculator is not being taught today. In addition, the College Board began to allow graphing calculators when students took the Calculus Advanced Placement Examination; beginning in 2000 students will be expected to bring along such tools to the test.

Mathematics Instruction

As recently as 1994, mathematicians and educators were still uncertain enough to voice concerns about the impact of the computer on the instruction of mathematics (Wilfe, 1994). Through the years, mathematics instruction has changed, based primarily upon the computer. From the early and crude computer-assisted instruction efforts, where instructional techniques were employed because they were easy to use rather than educationally sound

(Shank & Edelson, 1989), through software such as LOGO for the elementary school, through MAPLE for the university, to today's Web-based calculus courses offered by such outstanding universities as Oregon State and Texas A&M, the computer has affected the way mathematics is taught (Borba, 1995). At least one study reports problems in this ongoing process. Wenglinsky (1998) cites difficulties caused by how the computer is used, by an absence of equity in its availability (among the poorer schools), and by the uneven preparation of teachers in its implementation in their classrooms. Confirmation of the existence of these issues on a national scale can be seen in an analysis of the National Assessment of Educational Progress in mathematics data (Franks, Devaney, Katayama, Weerakkody, & Arnold, 1996).

Having teachers learn to use the computer is not an easy task to accomplish; it is a slow process and is full of plateaus when little or no progress is noted (Risku, 1996). Apparently, the potential is there: Despite the difficulties encountered in teachers' implementation of the computer in their instruction (Manoucherhri, 1999; Myrhe, 1998), Becker (1991) reported that the use of the computer in secondary schools more than doubled in only five years.

Reports of implementation of the computer into instruction include Hastings and Rossman's assessment (1994), which addresses the use of the computer in testing and placing students in a mathematics program. Hummel and Smit (1996) discuss both the development of new mathematics courses especially designed to use the computer and of self-study courses by distance. Similar activities in European schools are reported (Ponte, Nunes, & Veoso, 1991).

A somewhat higher order of use is delineated by programs using the computer to improve student ability to problem solve (Hersberger & Wheatley, 1989; Parish, Partner, & Whitaker, 1987). A more specific suggestion comes from Mathias (1998), who recommends that the computer is best used in an inquiry mode of instruction.

From a completely different perspective, it has been researched and thus suggested that the computer can be a significant tool in teaching mathematics to the learning disabled (Rapp & Gittinger, 1993).

Regardless of the method or intent of the use of the computer, and regardless of any difficulties that may be present, one can visit almost any school, elementary through college, and see rooms set aside for computers. Obviously, they are in use. They are a part of instruction.

But is anyone learning from them? Thomas, Tyrrell, and Bullock (1996) suggest not, writing that use of the computer is unlikely to result in a change in learning, until the personal and pedagogical philosophies of teachers change. Thus, those of us in mathematics teacher preparation need to take notice and instruct the pre-service teachers in how and why to use the computer! This may not be unlike the matter of having teachers in elementary schools use mathematical manipulatives (Tooke, 1992).

Learning Mathematics

A very comprehensive survey of the impact of the computer on the learning of mathematics was made in 1996. McCoy (1996) reviewed the research, and her findings were mixed; her recommendation was for more research in the area. However, a thoughtful reading of her report suggests that the number of findings with significant results seemed to be numerous at the elementary grades and diminished as the grade increased–to the point of very little, if any, significant results at the university level. This could easily be the logical consequence of either low-use levels or low research work as the grade level increases, and may be exactly why McCoy was calling for more research.

Certainly, from the perspective of a faculty member at an institution where distance education is a major part of the curriculum, this writer can confirm the previous statement, at least in one component: It is extremely difficult to recruit mathematics faculty to teach distance courses in upper-division mathematics. It is even more difficult to find upper-division Web-based courses for students (not exactly a call for "research this," but clearly a marketing possibility).

Wenglinsky (1998) states that, though research results are not clear and certainly not conclusive about the correlation between using the computer to teach mathematics and the learning of mathematics, one thing can be concluded from the research conducted: The computer is neither a cure-all nor a fad without impact.

CONCLUSION

McCoy and Wenglinsky may have presented the gist of this introduction. The computer has had an impact on all facets of mathematics education, even if the exact nature or degree of the impact is not certain. But it is obvious to all that the role of the computer will continue to grow, and, consequently, more research is needed to ensure correct and appropriate usage.

This volume has collected current writing from researchers and practitioners who have investigated or used the computer to expand mathematics education in some of the same areas just addressed.

Allen discusses the Texas A&M Calculus Web course mentioned earlier. Connell discusses the computer's role in teaching mathematics. More specifically, Stephens and Kovalina report on the computer's contribution to student success in learning intermediate algebra. Mayes reports on the computer's contribution to the teaching of algebra. Dugdale addresses the computer's role in student learning probability. Battista, Hannafin and Scott, and Liu and Cummings address the computer's role in students' learning of geometry.

Shotsberger discusses using the computer in the professional development of mathematics teachers. Wiest discusses the impact of computer technology on mathematics teaching and learning; and, finally, Maddux discusses the impact of the computer on perhaps the worst dread of mathematics educators: math anxiety.

REFERENCES

Becker, H. J. (1991). Mathematics and science uses of computers in American schools. *Journal of Computers in Mathematics and Science Teaching, 10*(4), 19-25.

Borba, M. C. (1995). Teaching mathematics: Computers in the classroom. *Clearing House, 68*(6), 333-334.

Cornu, B., & Ralston, A. (Eds.). (1992). The influence of computers and informatics on mathematics teaching. *Science and Technology Education Series, 44*, 134-140.

Cusco, A. A., & Goldenberg, E. P. (1996). A role for technology in mathematics education. *Journal of Education, 178*(2), 15-32.

Ediger, M. (1998). *Computers in the mathematics curriculum. Classroom teaching guide.* Kirksville, MO: Truman State University. ERIC Document Reproduction Services No. ED 421 991.

Franks, M. E., Devaney, T. A., Katayama, A. D., Weerakkody, G. J., & Arnold, C. L. (1996, April 8-12). *An analysis of NAEP trial state assessment data concerning the effects of computers on mathematics achievement.* Paper presented at the Annual Meeting of the American Educational Research Association, New York, NY.

Hastings, N. B., & Rossman, A. (1994). Workshop mathematics: Using new pedagogy and computers in introductory mathematics and statistics courses. Unpublished manuscript, Dickinson College, Carlisle, PA. ERIC Document Reproduction Services No. ED 418 839.

Hersberger, J., & Wheatley, G. (1989). Computers and gifted children: An effective mathematics program. *Gifted Child Quarterly, 33*(3), 106-109.

Hummel, H., & Smit, H. (1996). Higher mathematics education at a distance: The use of computers at the Open University of The Netherlands. *Journal of Computers in Mathematics and Science Teaching, 15*(3), 249-265.

Kaput, J. J. (1992). Technology and mathematics education. In D. A. Grouws (Ed.), *Handbook of research on mathematics teaching and learning* (pp. 515-556). Reston, VA: National Council of Teachers of Mathematics.

Knuth, D. E. (1994). Computer science and its relationship to mathematics. In J. Ewing (Ed.), *A century of mathematics* (pp. 285-288). Washington, D.C.: The Mathematical Association of America.

Manoucherhri, A. (1999). Computers and school mathematics reform: Implications for mathematics teacher education. *Journal of Computers in Mathematics and Science Teaching, 18*(1), 31-48.

Mathias, M. H. (1998). Mathematics, computers, and the real world. *Australian Mathematics Teacher, 44*(4), 2-6.

McCoy, L. (1996). Computer-based mathematics learning. *Journal of Research on Computing in Education, 28*(4), 438-460.

Myrhe, O. R. (1998). I think this will keep them busy: Computers in a teacher's thought and practice. *Journal of Technology and Teacher Education, 6*(2-3), 93-103.

National Council of Teachers of Mathematics (1987, November). *News Bulletin, 24*(2), 3.

Parish, C. R., Partner, B. E., & Whitaker, D. R. (1987, May-June). Mathematics and computers in the classroom: A symbiotic relationship. *School Science and Mathematics, 87*(5), 387-391.

Ponte, J. P., Nunes, F., & Veloso, E. (1991, July). *Using computers in teaching mathematics: A collection of case studies.* A product of Project Minerva. Projecto Minerva-DEFCUL, Faculdas de Ciencias de Lisboa, Campo Grande, 1700 Lisboa, Portugal.

Rapp, R. H., & Gittinger, D. J. (1993, November 14-17). *Using computers to accommodate learning disabled students in mathematics classes.* Paper presented at the Annual Conference of the League for Innovation in the Community College, Nashville, TN.

Risku, P. (1996). Mathematics teachers' approaches to computer-based instruction. *Scandinavian Journal of Educational Research, 40*(2), 137-159.

Sarmiento, J. (1997, June). *New technologies in mathematics.* Paper presented at Princeton University, NJ.

Shank, R., & Edelson, D. (1989). A role for AI in education: Using technology to reshape education. *Journal of Artificial Intelligence, 1*(2), 3-20.

Thomas, M., Tyrrell, J., & Bullock, J. (1996, April). Using computers in the mathematics classroom: The role of the teacher. *Mathematics Education Research Journal, 8*(1), 38-57.

Tooke, D. J. (1992, November). Why aren't manipulatives used in every upper elementary mathematics classroom? *The Middle School Journal, 24*(2), 61-62.

Waits, B. K., & Demana, F. (1998). Providing balance in the mathematical curriculum through appropriate use of hand-held technology. In Z.Usiskin (Ed.), *Developments in school mathematics education around the world* (pp. 275-286). Reston, VA: National Council of Teachers of Mathematics.

Wenglinsky, H. (1998). *Does it compute? The relationship between educational technology and student achievement in mathematics.* Princeton, New Jersey: Educational Testing Service (ERIC Document Reproduction Services No. ED 425 191).

Wilfe, H. S. (1994). The disk with the college education. In J. Ewing (Ed.), *A century of mathematics* (pp. 309-310). Washington, D.C.: The Mathematical Association of America.

Cleborne D. Maddux

Computers, Statistics, and the Culture of University Mathematics Education

SUMMARY. Most students dread taking a course in statistics. Negative attitudes toward statistics have many different causes. Education majors are not usually required to take an undergraduate statistics course, and practicing teachers thus do not appreciate the utility of theory in general or research in particular. Other causes are more related to the way statistics and mathematics courses are taught at the university level. Professors who believe that mathematically intense courses should serve primarily as a weeding-out process sometimes intentionally make their courses more difficult than they need to be, or institute high-handed, unjust policies designed to encourage dropouts. Undue emphases on rote memorization and calculation also contribute to the problem. Because of recent developments in distance education, such high-handed policies are doomed. As students have access to a variety of higher-education courses and programs, they will refuse to patronize traditional universities with policies such as those described above. Universities will be forced to adopt more rational policies. Computers can be used to relieve students from tedious, repetitive tasks and to create more and better opportunities to teach mathematics concepts and application. *[Article copies available for a fee from The Haworth Document Delivery Service: 1-800-342-9678. E-mail address: <getinfo@haworthpressinc. com> Website: <http://www.HaworthPress.com> © 2001 by The Haworth Press, Inc. All rights reserved.]*

CLEBORNE D. MADDUX is Professor, Department of Counseling and Educational Psychology, University of Nevada, Reno, NV 89557 (E-mail: maddux@unr.edu).

[Haworth co-indexing entry note]: "Computers, Statistics, and the Culture of University Mathematics Education." Maddux, Cleborne D. Co-published simultaneously in *Computers in the Schools* (The Haworth Press, Inc.) Vol. 17, No. 1/2, 2001, pp. 9-15; and: *Using Information Technology in Mathematics Education* (ed: D. James Tooke and Norma Henderson) The Haworth Press, Inc., 2001, pp. 9-15. Single or multiple copies of this article are available for a fee from The Haworth Document Delivery Service [1-800-342-9678, 9:00 a.m. - 5:00 p.m. (EST). E-mail address: getinfo@haworthpressinc.com].

KEYWORDS. Statistics, mathematics education, mathophobia, mathematics, distance learning

It is a gross understatement to say that most students dread taking a course in statistics. This is particularly true of students enrolled in university programs in the social sciences and the helping professions. Such students often regard the prospect of a first course in statistics with the same anticipation as that accorded an impending root canal or a visit to the state department of motor vehicles.

There are many reasons for the negative attitudes of education students toward courses in statistics. At the outset, I would make the point that university teacher training programs are partly to blame for this problem. An obvious contributing factor is that most undergraduate teacher education programs do not require *beginning* courses in statistics or research methods. Thus, practicing teachers usually have no foundation knowledge in statistics or research methods, no comprehension of what a statistics course might include, and no confidence in their own ability to successfully complete such a course.

More seriously, and partly because of this undergraduate omission, *many teachers do not believe research has a practical application,* and thus regard statistics not as an essential tool of their trade, but as simply another worthless requirement–yet another "hoop to jump through" on their journey to a master's degree. This lack of regard for research is partly due to the lack of regard for educational *theory.* It is a sad fact that teacher education programs across the country have been notoriously unsuccessful at convincing their undergraduate students that theory can be useful to them in their work in the classroom with students. This, in my opinion, is the single most universal and calamitous failure of U.S. teacher education programs and one of the most important reasons why public school teaching is held in low esteem and is not widely recognized as a full-fledged profession by members of our culture at large.

There is abundant evidence that teaching falls short of achieving professional status in the United States. Ernest Greenwood (1957) was a sociologist who carried out a classic study on the attributes of professions. He identified the existence of five characteristics of all American professions: (a) *existence and use of a systematic body of underlying theory,* (b) *recognition by the public of the expertise of the profession,* (c) *the sanction of the community,* (d) *existence of a regulative code of ethics,* and (e) *a professional culture.*

Later, Grace Graham (1969) evaluated the status of teaching with regard to each of the characteristics identified by Greenwood. She concluded that teaching has only two of the necessary five characteristics (a *professional culture* and a *regulative code of ethics*) and that teaching, therefore, does *not* meet the criteria for a profession. Graham commented at length concerning each of the characteristics and the status of each in teaching. She asserted that, while there *is* an underlying, systematic body of educational theory, most teachers, unlike physicians, lawyers, and other recognized professionals, *are not aware of it, do not value it, and do not base their practice upon it.*

Although Graham's conclusion that teaching was not a profession was arrived at more than 30 years ago, I believe she would come to the same conclusion today. Public education is undergoing the most sustained and vicious attack in modern history; politicians, business leaders, and almost everyone else believe that they know more about how education should be conducted than do educators; and many teachers and teacher groups advocate removal of teacher education from the curricula of colleges and universities and substitution of an apprentice program in public schools modeled after those used to train carpenters, plumbers, electricians, and other tradespeople.

Sadly, most teachers still do not value theory, and many actively and vocally denigrate its importance. I recently had an experience that illustrated the low opinion many teachers have concerning the value of theory. A colleague and I had just completed a manuscript about individuals with learning disabilities that was to be published as a book for teachers. When we sent the manuscript off to the publisher, we included a proposed subtitle: *A Theoretical Approach.* We were shocked when the publisher flatly refused to use this subtitle on the grounds that teachers would simply not purchase a book that mentioned theory so prominently, even if only in the subtitle!

With regard to attitudes toward research and statistics, if students do not perceive a use for theory, they are unlikely to believe that research is important, and they are equally unlikely to appreciate a requirement that they take one or more statistics courses, since statistics is a tool of research.

Another problem is that statistics is widely (and erroneously) believed to require mastery of difficult, advanced mathematics. This is unfortunate, not only because it is untrue, but also because many students have extremely negative attitudes toward mathematics. This problem prompted Papert (1980), an M.I.T. mathematician, computer scientist, and inventor of the Logo computer language, to observe that the entire Western world suffers from "mathophobia" (p. 38), an unreasonable and intense hatred and fear of mathematics. Papert went on to hypothesize that this negative attitude toward mathematics is caused by the way we teach it in schools:

> Our education culture gives mathematics learners scarce resources for making sense of what they are learning. As a result, our children are forced to follow the very worst model for learning mathematics. This is the model of rote learning, where material is treated as meaningless; it is a *dissociated* model. (p. 47)

Although Papert was referring to the teaching of mathematics in public schools, the passage might just as appropriately be applied to a great deal of the teaching of statistics in colleges and universities. For years, probably for generations, statistics professors have emphasized the rote mathematical *calculations* central to statistics, and have neglected to teach the *application* of statistics and statistical *concepts*. As a result, legions of students have graduated from master's degree programs and even from doctoral programs in education possessing only rudimentary ability to apply statistical formulae to calculate *t*-tests, analyses of variance, chi-squares, etc., on demand, but without the slightest idea of when to apply each test, or how to interpret the results of such tests carried out with actual research data.

This emphasis on calculation rather than on application and interpretation is exactly analogous to the situation in the public schools with regard to the teaching of long division. Achievement test after test has shown that the public school's single-minded emphasis on long division calculation and obtaining the correct answers on worksheets listing rows of division problems, while unsuccessful in itself, has resulted in generations of children who hate and fear mathematics in general and long division in particular, and who are unable to solve even simple word problems that do not specify the proper mathematical process. *In short, emphasizing calculation and neglecting concepts and interpretation alienate learners and do not result in the ability to apply mathematics to problems encountered in life outside the mathematics classroom.* The same can be said of statistics teaching in which the main approach is assignment of lengthy textbook homework exercises at the end of each chapter. Any one assignment tends to be solely directed at one specific statistical test, and does nothing to help students learn to select the proper test, organize the data, or perform any of the other myriad tasks that researchers carry out in the real world of research.

Why have statistics professors chosen to emphasize rote calculation and neglect concepts, application, and interpretation? One reason is that, before electronic calculators and computers, they had little choice. Statistics had to be hand-calculated, and the only way to learn to do so was through laborious and frequent drill and practice. However, electronic calculators have been common for at least thirty years, and personal computers and advanced statistical programs have been widely available for more than ten years. Why, then, have professors continued to take a rote approach to the teaching of statistics?

Tradition plays a bit of a role, but there is another, more insidious explanation. *Many statistics professors have internalized a particularly sinister and all-too-common subculture of university mathematics teaching.* This subculture is one in which student failure is worn as a badge of honor, "rigor" is a euphemism for poor teaching, and "weeding out" is the near-sacred, primary professorial mission.

I would hasten to point out that by no means do I imply that all mathematics professors are part of this subculture of defeat. It goes without saying that there are many excellent teachers in mathematics departments. However, enough mathematics professors fit the portrait described to make it readily recognizable by anyone who has taken a few mathematics courses at a college or university. Indeed, some of my statistics students have told me that they so dreaded taking a mathematics-related course that they were physically ill during the week before the class began!

I have no idea what can be done about the problem of professors who delight in students' failure. Thank goodness such professors are in the minority. Yet even a few such instructors are exposed to many students and can do substantial harm over the course of an entire career. During my own tenure as a university student, during the last 25 years of teaching at five different universities, and as the parent of two university graduates, I have become aware of many nearly unbelievable examples of this general phenomenon. I could list many such examples, but a few illustrative cases come to mind.

I especially remember the mathematics professor at my own institution who directed his 20 students to pick up their weekly homework assignments from an envelope taped to his office door. However, each week, the envelope contained only 10 copies. When students complained that there were not enough to go around, he suggested that they should plan to pick up the assignment earlier.

I remember, too, the mathematics professor whose idea of teaching was to walk into class 15 minutes late, write the answers to the homework problems on the board, mutter "Any questions?" and stalk out of class before anyone had a chance to respond.

I remember the professor who routinely spent the first three weeks of each fall term consulting in a foreign country, and who never bothered to enlist someone to cover his classes, or even to inform students that classes were cancelled. No message concerning cancellation was ever printed on the classroom chalkboard. The professor simply did not show up for the first three weeks of classes.

I suspect that these are examples of a host of outdated, but common, attitudes and policies toward university students–attitudes and policies that are going to come to a sudden and forcible halt in the very near future. Although it is a topic for another day, I say this because I believe that we are at a turning point in American higher education. That turning point is being forced upon us by the Internet, the World Wide Web, and by all the manifestations of distance education that are rapidly establishing a student "buyer's market."

Throughout modern times, universities have been able to treat students in outrageously high-handed ways because most students had very little choice.

For a variety of economic and personal reasons, most students were unable or unwilling to attend any institution other than the one that was geographically closest to their home.

That kind of "seller's market" is coming to an end. Private entrepreneurs as well as established, "tier-one research institutions" are falling all over one another in their haste to develop and market entire bachelor's, master's, and doctoral degrees over the Internet. Hardly a week goes by that I do not receive several brochures from such public and private institutions. I have one such brochure on my desk as I write this article. This brochure states that all previous education and work experience will count toward bachelor's, master's, or doctoral degrees, most of which can be earned online in 12 months or less and completely without mandatory attendance in structured classroom settings.

What the availability of this and many other similar programs means is that *higher education students of the future will have choices.* Competition, for good or for ill, is arriving on campus. In the face of that competition, I do not believe students will continue to pay inflated parking fees for the right to hunt for non-existent parking, and I do not believe that they will be willing to continue to pay hard-earned cash for the right to sit in a mathematics or statistics class that has been intentionally designed to fail a substantial, prede-termined percentage of students, or one in which the instructor boasts pride-fully about the number of *F*s awarded the previous semester.

These and many other such ill-advised attitudes and practices are doomed to extinction in the face of escalating higher education competition. Although I do *not* believe that the overall effect of such distance education programs on the quality of higher education will be positive, I *do* believe that they may serve to spark a great many needed reforms in the way university faculty members and administrators treat students.

But let us return to the topic of statistics teaching and address how to improve it. Although I have no magic solution to change the attitudes of those who believe their proper role is to make mathematics or statistics harder than they need to be, I believe that computers can be of enormous use to those of us who hope to help our students learn to apply mathematical and statistical concepts to real problems outside the classroom.

Although I do not believe that it is advisable to teach statistics in such a way that students are *never* asked to perform any mathematical calculations, computers can obviously relieve our students from the necessity of mindless, repetitive drill and practice of tedious, rote calculations. This can free up time to teach students how to select the proper analysis, how to prepare data for entry into statistical programs, how to read and interpret printouts, and how to write up research reports for submission to refereed journals.

In fact, I believe that computers have the potential to revolutionize education in general, and the teaching of mathematics and mathematically-related topics in particular. Computers have many potential roles to play in education, and I believe one of the most important is the ability to relieve students from tedious, repetitive tasks and create more and better opportunities to teach mathematics concepts and applications.

This volume is dedicated to the use of computers in teaching mathematics-related subjects. These articles do an excellent job of showing how computers have the potential to reverse the trend toward mathophobia among our students.

REFERENCES

Graham, G. (1969). *The public school in the new society.* New York: Harper & Row.

Greenwood, E. (1957, July). Attributes of a profession. *Social Work, 2,* 45-55.

Papert, S. (1980). *Mindstorms: Children, computers, and powerful ideas.* New York: Basic Books.

G. Donald Allen

Online Calculus:
The Course and Survey Results

SUMMARY. The details of the inception, development, and implementation for an online mathematics course, Engineering Calculus I, are presented. Several of the major challenges in the design and layout of the content are given special attention and their contrast with textbook structure is noted. With numerous surveys given throughout the course, we were able to continuously monitor student reaction to the venue and comprehension of materials. In pedagogy, several observations were made about how students learn from Web-based materials. These have formed a key component during the continuing development stage of this project. *[Article copies available for a fee from The Haworth Document Delivery Service: 1-800-342-9678. E-mail address: <getinfo@haworthpressinc. com> Website: <http://www.HaworthPress.com>© 2001 by The Haworth Press, Inc. All rights reserved.]*

KEYWORDS. Calculus, Web-based, interactive learning, facilitator Web-assisted, computer algebra system, Maple

THE WEBCALC PROJECT–INCEPTION

The WebCalC Project, whose team members are Michael Stecher, Philip B. Yasskin, and this author, is a complete "course-in-a-box." The course is

G. DONALD ALLEN is Professor, Department of Mathematics, Texas A & M University, College Station, TX 77843 (E-mail: dallen@math.tamu.edu).

[Haworth co-indexing entry note]: "Online Calculus: The Course and Survey Results." Allen, G. Donald. Co-published simultaneously in *Computers in the Schools* (The Haworth Press, Inc.) Vol. 17, No. 1/2, 2001, pp. 17-30; and: *Using Information Technology in Mathematics Education* (ed: D. James Tooke and Norma Henderson) The Haworth Press, Inc., 2001, pp. 17-30. Single or multiple copies of this article are available for a fee from The Haworth Document Delivery Service [1-800-342-9678, 9:00 a.m. - 5:00 p.m. (EST). E-mail address: getinfo@haworthpressinc.com].

17

Calculus I, a typical engineering calculus course, equivalent to an AP calculus (but without a graphing calculator component). The inception of this online course was in the spring of 1997. Several months of discussion ensued, particularly over the two most essential points.

1. What should be the method/software for delivery?
2. What does an online course need to be?

Dozens of other issues, from the file management system to the creation of graphics, were ignored at first, though each eventually required careful attention. The second question has a simple solution summarized as follows:

> At a minimum the complete online course must do everything a book does. To succeed, it must do very much more. Developers should look for computer-assisted teaching devices that the classroom teacher cannot match. (Allen, Stecher, & Yasskin, 1998a, p. 62)

A rather simple statement to make; meeting such a requirement, however, it is difficult. For one thing, what is the "more" alluded to above? For another, the continuous and rather steep learning curve of Web technology, on which most of us find ourselves, offers its own challenges. It would be easier if the technology were relatively static, like book technology, but Web technology and its correspondent hardware technology are changing at an ever-increasing rate. Therefore, decisions made one day may not be valid the next. It is daunting to realize that a whole online text effort may have to be scratched and recast in some new format, possibly even before completion.

However, experience has shown that the online course must be every bit as complete as a book, and more so. For example, answers to sample questions should contain more detail than the typical textbook. Mathematics textbooks, by the way, are not written for self-study and fail miserably if so used without supplementary materials. Even so, they are used exactly this way, and without supplements, in many distance-education efforts. The online course, however, must be written exactly for the purpose of use 'at a distance,' whether that is distance education or self-study.

For content creation, we decided on what could be termed an "onion" paradigm, from which we would begin with a core text, then modify and add to it as time, technology, and skills permitted. In essence, we built the course a layer at a time.

While the second question can be dismissed with lofty goals, the first question, a significant operational detail, remains: What presentation software should be used? Conventional wisdom dictated that HTML-only is not only the reasonable choice, but also the only choice. While this prescription is adequate to convey information for many courses, we questioned whether

HTML-only was good enough for a mathematics-intensive course. The decision did not come easily. It was essential to determine exactly what the student should see. With this in mind, we prioritized those features we assessed as most important. Overall, the course should have an economical, functional, and attractive design (Lynch & Horton, 1999). Precisely, the course should have:

1. Perfect mathematical typography–math should look like math;
2. Great color and graphics;
3. Interactive quizzes and exams;
4. Internet links; fast downloads;
5. Symbolic mathematics capabilities;
6. Complete solutions to examples and exercises;
7. Question-answer notes;
8. Animation and Java;
9. Sound and video;
10. Years of testing.

The logic for most of these requirements has been discussed in some detail (Allen, Stecher, & Yasskin, 1998a & 1998b). Consequently, we make here only a few supplementary points. First of all, students are accustomed to the highest quality software, whether from the viewpoint of a word processor or spreadsheet program, or from their vast experience with computer games. Therefore, the key point of excellent mathematical typography is in the interest of student expectations as well as student comprehension, particularly the beginning student enrolled in calculus. While mathematics instructors have little difficulty reading mathematics in almost any form, the beginner does have trouble. For these students, the presentation must look as perfectly textbook-like as possible. This decision clearly limited available choices. The most direct way to achieve good mathematical typography is through GIF images created by programs such as LATEX2HTML or, say, through HTML export of an MSWord document. Both procedures generate mathematics of good appearance, but neither prints well. In addition, resulting file sizes tend to be large. This means files download slowly, and fast download times are essential. Student interest and attention fade if lengthy downloads persist. (Roughly, a consistent 20 seconds download time per page is near the maximum time students will accept.) A couple of other methods are available, but they each suffer one or more defects. Robson (in press) discusses object-oriented course design for Web-based authoring to alleviate some of the extraordinary complexity for developing such a project.

The last three requirements (items 8 through 10) were considered only after the project was underway. Indeed, we have added a few animations (animated GIF images) and Java applets. Such devices have not yet become *essential* course features, although students do like them. Sound and video were not feasible when we began, but thanks to new, highly compressed streaming software, they now are. During the fall of 2000 we will begin to add narrated slide shows to the course in select places (Levine, 1999). The last point about testing was unrecognized only after testing began, and we began to understand how complex this aspect of the project was. Briefly, it is important to understand that learning how to use an online course is a separate skill altogether. Hypothetically, even the perfect online course could fail its field tests if the instructor is not given directions on how to administer it. Further details of this fact, vis-à-vis WebCalC, are described later.

Our software decision was to use Scientific Notebook, a TeX typesetting engine, with a built-in Internet browser and a computer algebra system (Maple). TeX is the de facto standard for mathematical typesetting today. It has been used for more than 20 years and produces the best-appearing mathematics of any competing method. This generally makes for very small files, normally 5 to 10 kilobytes, excluding graphics. The browser part of the program builds the page from the TeX codes when it arrives at the client's computer. With the Maple kernel, which is also menu driven, most students can answer many of the "what if" type questions that typically come up or produce great graphics with hardly more than the click of a button. Finally, it is exceptionally easy to typeset mathematics using Scientific Notebook, as that feature is also menu driven. Often portions of the material must be rewritten as the developer notes how students respond to it. Remember that *without the lecture, there is little opportunity to clarify what has been written.*

Though graphics in abundance have emerged in mathematical textbooks only in the last half of the twentieth century, they are today regarded as essential. In that connection, students, having experience with cutting-edge programs, are again the experts. The graphics must be excellent. We construct ours using a variety of methods, the most dominant being Corel Draw to generate Windows metafiles (i.e., vector graphics). These Windows metafiles (WMF) are superior to GIF or JPEG images in almost every way except possibly file size. Most of our graphics are 30 kilobytes or less. They are in full color-as needed-and are in quality at the level of modern textbooks. As line speeds increase over the next few years, we look forward to generating graphics that are ever more sophisticated and interesting.

THE WEBCALC PROJECT-
MORE FEATURES AND REMARKS

Basically, our course does meet the requirements outlined in items 1 through 10. Importantly, the lean and lively text pages (Douglas, 1986) are written to cover all the essential points needed for the topic at hand. Neither lengthy introductions nor long summaries are included. It is curious that, in recent reviews of WebCalC, one reviewer noted this as a negative feature, but further noted that the preponderance of students don't read introductions and long summaries anyway. We've noted that, just as in the traditional course, students tend to begin with the problems and only read the text as needed. Without the lecture wherein students see examples worked and theory explained, students of online courses have less instruction than their traditional counterparts. Therefore, proceeding directly to the problems is a riskier strategy. On this basis, one fundamental observation we have noted is that *learning from an online course is task-oriented.* Simply assigning chapters to be read and exercises to be worked is too abstract or vague to be effective. It is important to give precise tasks to the student on a daily basis. This can be accomplished through quizzes and worksheets. To drive this point home, every instructor, following the traditional method of teaching, is substantially the *metronome* for his or her course. The lecture sets the tasks, and the students respond by working the problems or reading specific material. This facet of traditional teaching needs an analogue in the online course. While there may be better methods yet undiscovered, for now the quizzes and worksheets seem to be working.

The online course requires a number of other features, and WebCalC has them.

1. Randomized quizzes
2. Homework assignments with complete answers
3. A consistent, attractive look and feel that is artistic
4. Notes (pages within pages)
5. Symbolic mathematics functionality

Many of these requirements can be lumped under the rubric of *interactivity.* It is important for the course to communicate with the student by asking questions and providing answers, though not at the same time. This is accomplished through a variety of techniques. One is the randomized quiz. These quizzes are generated on the fly and provide different questions (over the same topic) each time the button is selected. For example, think of a problem involving solving a linear equation where, each time the button is chosen, a different linear equation appears. When answers to the question or questions have been selected, the student receives immediate feedback. The course also contains blue ball quizzes, so named because of the blue balls (buttons) next to the multiple-choice answers. These quizzes are carefully crafted with feedback provided with each wrong answer and with explanation given for the correct answer. (Some students just choose buttons until the correct answer is found.)

The online course should have a nonlinear structure. Thus, the student can be maneuvered about as the links and hyperlinks determine. One particularly nonlinear, useful feature is the pop-up note. For example, when explaining the solution to a problem, it may be necessary to invoke a previous concept, say the Pythagorean theorem. At this point, a button is inserted; and the student, by choosing it, automatically sees the needed information. This is just one use of pop-up notes. Pop-up notes can be used for making definitions, for reinforcing a concept, and for supplying a relevant graphic. They can also be used to enhance text information by supplying additional facts that would clutter the main page. Such notes are downloaded with the page and appear only if selected. Our most frequent use of the pop-up note has been to supply answers to exercises. Next to each exercise problem there is an answer button which the student chooses to view the answer. Overall, such notes give a definite appearance of interactivity. Links and hyperlinks are used for larger notes, for going to the next page, and for additional information of a nature inappropriate for a small pop-up box.

WebCalC has a few animated GIF images that students like, as well as a few fully interactive Java applets. One applet plots a user-supplied function and then displays a secant line evolving into the tangent line. As interesting as this is, students don't seem to use it. We believe that some specific task must be associated with the applet for it to achieve a more prominent place within the course. Some Java applets tend to download and launch slowly. A delay of a minute or more is not uncommon.

THE WEBCALC PROJECT–USING THE COURSE

In an ideal world, pre-selecting students is the rule. Hence the question: If our world was ideal, who should and who should not take an online mathematics course? Our first choice would be to select students that are strongly motivated (self-starters), intellectually mature, home-schooled, or individuals with a handicap. Our last choices would be students that have low mathematics pretest scores, are generally unmotivated, or need the structured classroom environment. At this point of online course development, the temptation to view these courses or require them to be suitable for all students is at best misguided, at worst a recipe for disaster. At minimum, a key ingredient to the success of the online course is a clientele that have very good reading skills. A very few (about 3 out of 150) of our WebCalC students decided that online calculus was not for them. Every effort was made to transfer them to a traditional section.

By the beginning of the spring 1998 term, we had taught WebCalC to a total of five sections of 25-35 students each. However, typical of the total four sections offered per semester, two were scheduled for freshmen mathematics majors and were to be delivered via the traditional lecture mode and the other two were scheduled for physics, chemistry, genetics, and geoscience majors and were delivered online. In effect, the two sections for WebCalC had all the freshman chemistry, physics, genetics, and geoscience majors. Thus it was not the ideal. Although we did not survey students about their course selection at the time, we did survey them extensively as the terms progressed. The mode of teaching, termed the *facilitator mode*, has several essential factors in common with the traditional mode. In the facilitator mode, the instructor is in the classroom with the students during a regularly scheduled period. The students, however, sit at a computer terminal and learn from the online course. The teacher/facilitator (a) answers questions in situ, (b) assigns/collects/grades homework, (c) establishes the pace, (d) creates/grades exams and quizzes, and (e) holds office hours. In brief, the facilitator performs all the traditional functions except to deliver a lecture. The time traditionally spent lecturing is spent helping individual students with individual problems, usually to explain a worked example or to explain how to answer homework questions. The facilitator is generally occupied with student questions throughout the period.

What we have noticed is that students group in pairs or sometimes trios to study/learn together. Called *spontaneous group learning*, this remarkable phenomenon differs from other group learning since it is not contrived (Gantt, 1998; Jensen & Sandlin, 1991). Other students work strictly alone. Many of them never even ask a question. Most groups formed within a few weeks; other individuals migrated from one group to another. While we have no precise data about the results, our impression is that well over half the class worked in some loosely or tightly formed group, and that the grades of these students were higher on average as compared with "loners." That the more socialized students do better seems to be true, but there is a law of diminishing returns: Too much of this good thing seems to decrease grades.

We have yet to discuss the true distance-education mode, where there is no regular meeting time or place and where most communication is *not* face-to-face. In one sense, that is fortunate. Our experience with online courses has suggested that going to the full or true distance mode may have proven difficult had it been attempted earlier. The single most difficult question remaining is how to give proper and timely mathematical help to students as it is needed. Other documented problems include student isolation and frustration. The first, true distance-education version of WebCalC was offered during the fall 1999 term. We are mindful that students completing Calculus I must then proceed to a traditionally formatted Calculus II. Therefore, students must be provided first quality education from Web-CalC.

Some believe that because of technology, the very nature of the course, the syllabus itself, should change. This alters the burden of online course development. However, before massive changes in the course syllabus are implemented in synchrony with this philosophy, online courses should first satisfy the "proof of concept" of any new design with respect to the traditional syllabus.

THE WEBCALC PROJECT–SURVEY RESULTS

The measurement of course efficacy for the traditional course has been long abandoned. Typically, students are surveyed about the instructor with a sampling of questions about the course. Traditional pedagogy is old, time-tested, and well understood. On the other hand, online courses seem to need to prove themselves to a higher standard. College and high school administrators are anxious to join the new age of online courseware but are nervous about the quality of the product. Perhaps trust for their teaching colleagues is not what it could be, and perhaps they fully comprehend the difficulties presented by this new medium; but surely they desire that online courses meet a level of quality consistent with their institution.

The developers of WebCalC are conscious as well about the quality of their new product. They typically test the quality using a panoply of surveying instruments. The most basic measures of course efficacy are surveys that measure students' impressions and studies that measure students' grades in subsequent courses. The surveys taken for this study do not follow precise standards of statistical sampling (Hall, Pilant, & Strader, 1999) but are rather preference surveys given and collected during a class period, four times per semester. The randomness of the sample was addressed earlier.

Since this course has been taught just five terms, data have not convincingly established what we hope is true; namely, that online mathematics taught in the facilitator mode is competitive with, or better than, the traditional mode. However, if results are merely comparable after only five semesters of use, this marks a positive indicator of the potential of Web technology as the medium of the future.

A number of questions were posed to students; many of them routine in nature, many key questions of great importance to the project. Responses were on a scale from "Strongly Disagree" to "Strongly Agree." The more important questions and responses are displayed in Tables 1 through 8. Less significant questions to the reader are those involving material comprehension. Example: *I understood the material on inverse trigonometric functions well.* Other questions were included to measure students' "comfort level." Example: *I am very worried about the upcoming third exam.* Finally, some of the same questions were posed in alternate ways: for example, *The level of the material in the course was too advanced*; or: *I found that chapters adequately explained the material.* These questions are similar to the question in Table 1.

The mild "disagree" response indicates the normal confusion and difficulty that many students have with a beginning college mathematics course. Overall, the surveys indicate that the level and pace of the material are acceptable, at least to the students. In Table 2, we inquired about the perceived challenge of learning over the Web.

Should these responses be considered surprising? Not at all, because the student is doing most of the work in the course. No longer is the atmosphere passive, where the disinterested student can simply let a lecture flow around but not through his/her awareness. Students come to the lab to work. In Table 3, student study location habits are addressed. Do students work at home or in class?

TABLE 1. Question: I Found the Material Paced Evenly, with Just Enough Explanation Given

Term	Survey	Strongly Disagree	Disagree	Neutral	Agree	Strongly Agree
Fall 1998	1	2	5	6	13	3
	2	1	5	10	10	2
	3	1	8	9	14	0
	4	1	5	10	10	1
Spring 1999	1	1	4	10	13	2
	2	2	7	4	14	5

TABLE 2. Question: I Found Learning from the Web Far More Challenging Than I Ever Imagined

Term	Survey	Strongly Disagree	Disagree	Neutral	Agree	Strongly Agree
Fall 1998	1	2	6	12	6	3
	2	0	5	9	6	5
	3	1	1	9	4	8
Spring 1999	1	3	6	7	7	7
	2	4	7	3	12	6

The responses are generally neutral overall and are well balanced. The indicator points to possible results for the distance mode, which was attempted in the fall 1999 semester. It also demonstrates that many students feel that they accomplish much in the class period and can supplement their learning in the open computing laboratories throughout campus. The results presented in Table 4 reveal that the power of the software package is underutilized.

The responses in Table 4 were a little puzzling. Scientific Notebook, which functions as the Internet browser, has a powerful menu-driven Maple engine. It is very, very simple to learn and to use. Why didn't students use it? Probably because it was not emphasized in the classroom. Several students used graphing calculators to help them resolve questions. A few of the academically stronger students actually taught themselves how to use the Maple engine. However, most students did not venture far from the path of assigned work. Note that the use of a CAS (computer algebra system) in the classroom is not well understood, even after a decade of attempts (Allen, Herod,

TABLE 3. Question: I Preferred to Work on the Course at My Home Over the Net

Term	Survey	Strongly Disagree	Disagree	Neutral	Agree	Strongly Agree
Fall 1998	1	2	5	14	7	3
	2	1	9	7	7	1
	3	3	7	8	5	0
Spring 1999	1	5	6	10	9	0
	2	5	9	8	9	1

TABLE 4. Question: I Used the Symbolic Features of Scientific Notebook to Experiment or Check Mathematics Often

Term	Survey	Strongly Disagree	Disagree	Neutral	Agree	Strongly Agree
Fall 1998	1	8	8	5	3	6
	3	6	8	4	4	1
	4	5	10	2	8	3
Spring 1999	1	6	11	5	8	0
	2	14	8	8	2	0

Holmes, Ervin, Lopez, Marlin, Meade, & Sanchez, 1999). In Table 5 the comfort level of students with respect to live help is addressed.

The responses in Table 5 were gratifying, but, more importantly, illustrate the effectiveness of using the facilitator mode. Note that students did not often come to my office for help. These responses are generated from in-class help, as it was requested. The strong "agree" responses indicate that even the loner-type students seemed to feel that adequate help was available, although they may never have requested it.

Table 6 illustrates verification of what was observed–many students believe that they are actively working with others. The responses to this question would have been the most surprising of all, had it not been observed directly. From Table 6 it is clear that, for a majority of students, the learning process was collaborative. Called *spontaneous group learning*, this learning mode was attractive to many students and is radically different from the traditional lecture mode. Oftentimes questions would be asked by a small group of students. Additionally, the stronger students found themselves helping weaker ones, though I am not convinced of the efficacy of this. The principal

TABLE 5. Question: Adequate Help Is Available in Class or Elsewhere When I Need It

Term	Survey	Strongly Disagree	Disagree	Neutral	Agree	Strongly Agree
Fall 1998	1	1	2	0	19	8
	3	0	3	3	14	3
	4	0	3	7	14	3
Spring 1999	1	0	2	8	14	5
	2	1	3	4	16	8

TABLE 6. Question: I Regularly Work on Class Materials with Other Students

Term	Survey	Strongly Disagree	Disagree	Neutral	Agree	Strongly Agree
Fall 1998	3	0	6	3	8	6
	4	1	5	4	14	5
Spring 1999	2	3	6	6	12	5

reason is that the stronger student, though very capable, may not be in tune with the real problems of the weaker student. He/she may give help that results in further confusion. The most tangible example of this, which may be peculiar to mathematics instruction, is to notice the better student showing the weaker one a special way of solving a problem that is apparently quite different from the methods being taught. Indeed, the facilitator mode of teaching may work best when the facilitator is a highly experienced teacher well grounded in all the methods and techniques of the subject. This point serves to emphasize that the teaching component has by no means been diminished by the online format; rather, it has been enhanced.

It seems that learning in this mode is partly a social process. What is even more fascinating is that the students in the spontaneously formed groups tended to perform better on examinations, though these were administered in the traditional way, with no student interaction permitted. However, this has not yet been measured from grade results data. Finally, we noted that these spontaneous groups were somewhat self-selecting. Most groups consisted of students relatively similar in ability. With this group learning process in place, the lab is naturally less than a quiet and contemplative environment. Do the students perceive it that way? (See Table 7.)

Even though the lab atmosphere, with all the conversation, would be distracting in a traditional sense, most students did not regard it as so. We could have insisted on strict independence of work and silence in the lab, leaving only the hum of CPU cooling fans, but we would have lost noticing the remarkable feature of social learning in a task-oriented environment. The last question, "Would the student take another online course?" provided important feedback, from which several interpretations are possible (see Table 8).

The responses to the question in Table 8 were remarkably consistent over all semesters. They reflect the fact that the class had to work hard, that they believed they had learned the material, but that they were not yet ready to

TABLE 7. Question: The Lab Atmosphere Is Distracting

Term	Survey	Strongly Disagree	Disagree	Neutral	Agree	Strongly Agree
Fall 1998	3	0	10	6	4	3
Spring 1999	2	5	14	9	3	1

TABLE 8. Question: If the Next Calculus Course Were Available Online in the Same Format, I Would Take It

Term	Survey	Strongly Disagree	Disagree	Neutral	Agree	Strongly Agree
Fall 1998	4	4	5	7	5	2
Spring 1999	3	5	6	7	7	2

embrace online courseware with full vigor. These students, mostly freshmen, were able to learn calculus in a way entirely new to them, and this is significant.

CONCLUSIONS

WebCalC is a successful online project, principally because it works. By this, we mean that students can learn Calculus I online, perform successfully on traditional examinations, and perform well in successive calculus courses. Moreover:

1. Students adapt to online learning without difficulty.
2. Students learn to "get to work" right away, making efficient use of their time.
3. Students like the task-oriented environment.
4. Students can work together with positive results.

Some course revisions, although not many, have been required. More enhancements will be added as time allows. Precisely, we envision enhancements to include greater student friendliness, increased task orientation, increased ease of navigation, availability of sound and video slide shows, and faculty bulletproofing.

Student solicitousness and faculty bulletproofing are particularly important until the preponderance of instructors have experience with online courseware. There are numerous pedagogical techniques that must be observed to allow the online course to work. At this juncture in online education, many instructors, if given an online course teaching assignment, will treat it as some modification of what they already know. Administering the course on the basis of traditional face-to-face teaching may lead to unsatisfactory results.

However, it will be important to measure the "value-added" component, as enhancements come at an increasingly high cost. High production costs, high development skills, and high production time schedules are all deterrents to further online course development. For example, a professional quality Java applet may require several months of programming effort. Commercial costs can run into hundreds of thousands of dollars. Then, the applet must be tested for efficacy. In contrast, textbook technology is well tested and well understood, and accurate methods exist for estimating production costs.

It is significant that no doubt many false starts and many revisions to online pedagogical designs will be needed to perfect this mode of education. If costs are excessive, the funding may not be forthcoming, certainly from public institutions.

REFERENCES

Allen, G. D., Stecher, M. S., & Yasskin, P. (1998a, November/December). The Web-based mathematics course: A survey of the required features for an on-line math course and experiences in teaching one. *Syllabus Magazine*, pp. 62-65.

Allen, G. D., Stecher, M. S., & Yasskin, P. (1998b). WebCalC I: A description of the WebCalC project, its history and features. *Proceedings of the Eleventh ICTCM Conference* (pp. 231-238). New Orleans, LA: Addison-Wesley-Longman.

Allen, G. D., Herod, J., Holmes, M., Ervin, V., Lopez, R. L., Marlin, J., Meade, D., & Sanchez, D. (1999). Strategies and guidelines for using a computer algebra system in the classroom. *International Journal of Engineering Education, 15*(6), 112-134.

Bogley, W. A., Dorbolo, J., Robson, R. O., & Sechrest, J. A. (1996). New pedagogies and tools for Web-based calculus. In H. Maurer (Ed.), *Proceedings of the AACE WebNet96 Conference* [Online]. Available: *http://iq.orst.edu/papers/WebNet96*

Bogley, W. A., Dorbolo, J., Robert O., Robson, R. O., & Sechrest, J. A. (1998). Pedagogic innovation in Web-based instruction. In P. Bogacki, E. D. Fife, & L. Husch (Eds.), *Proceedings of the Ninth ICTCM Conference* (pp. 421-425). Reading, MA: Addison-Wesley. [Online]. Available: *http://iq.orst.edu/papers/ICTCM96.html*

Douglas, R. G., (Ed.). (1986). *Toward a lean and lively calculus.* (MAA Notes, No. 6). Washington, DC: Mathematical Association of America.

Hall, R. J., Pilant, M. S., & Strader, R. A. (1999, March 2). The impact of Web-based instruction on performance in an applied statistics course. In R. Robson (Ed.), *Proceedings of the International Conference on Mathematics/Science Education and Technology (M/SET)* (pp. 334-339). San Antonio, TX: Association for the Advancement of Computing in Education.

Levine, L. E. (1999, Summer). Mathematical slide shows. *The Math/Science-Online Newsletter.* [Online]. Available: *http://www.math.tamu.edu/~webcalc/newsletter_99b. html*

Lynch, P. J., & Horton S. (1999). *Web style guide.* New Haven, CT: Yale University Press.

Robson, R. (in press). Object-oriented instructional design and Web-based authoring *Journal of Interactive Learning Research.*

Paul G. Shotsberger

Changing Mathematics Teaching Through Web-Based Professional Development

SUMMARY. This article summarizes efforts of the University of North Carolina at Wilmington to deliver mathematics teacher professional development using a Web site called INSTRUCT. The site is intended to assist middle and high school mathematics instructors in implementing the National Council of Teachers of Mathematics professional standards for teaching mathematics (NCTM, 1991). Results from the project are presented along with a discussion of the benefits and limitations of this type of training. Three potential benefits associated with Web-based professional development have emerged from three years of experience with INSTRUCT: consistent opportunities for reflection and sharing; a shortened cycle for training, implementation, and evaluation; and just-in-time support for teacher change. *[Article copies available for a fee from The Haworth Document Delivery Service: 1-800-342-9678. E-mail address: <getinfo@haworthpressinc.com> Website: <http://www.HaworthPress. com> © 2001 by The Haworth Press, Inc. All rights reserved].*

KEYWORDS. Web, professional development, mathematics, in-service, just-in-time, NCTM, standards, teacher change

PAUL G. SHOTSBERGER is Associate Professor, Mathematics and Statistics Department, The University of North Carolina at Wilmington, 601 South College Road, Wilmington, NC 28403-3297 (E-mail: shotsbergerp@uncwil.edu).

Primary funding for this project has come from the Eisenhower Professional Development Higher Education grant program, administered by the North Carolina Mathematics and Science Education Network.

[Haworth co-indexing entry note]: "Changing Mathematics Teaching Through Web-Based Professional Development." Shotsberger, Paul G. Co-published simultaneously in *Computers in the Schools* (The Haworth Press, Inc.) Vol. 17, No. 1/2, 2001, pp. 31-39; and: *Using Information Technology in Mathematics Education* (ed: D. James Tooke and Norma Henderson) The Haworth Press, Inc., 2001, pp. 31-39. Single or multiple copies of this article are available for a fee from The Haworth Document Delivery Service [1-800-342-9678, 9:00 a.m. - 5:00 p.m. (EST). E-mail address: getinfo@haworthpressinc.com].

31

Use of the Internet for just-in-time professional development and support, once the primary domain of the business world, is beginning to have an influence on the education community (Owston, 1998; Romiszowski, 1997). In 1996, the Department of Mathematics and Statistics at The University of North Carolina at Wilmington (UNCW) recognized the potential of the Web for in-service teacher training and developed a site called INSTRUCT, which stands for *I*mplementing the *N*CTM *S*chool *T*eaching *R*ecommendations *U*sing *C*ollaborative *T*elecommunications (http://instruct.cms.uncwil.edu). INSTRUCT is intended to assist middle and high school mathematics instructors in implementing the National Council of Teachers of Mathematics (1991) *Professional Standards for Teaching Mathematics* by providing them online reading materials and the opportunity for reflection through dialogue with colleagues.

During the fall semester of 1996, the INSTRUCT site was piloted with four mathematics teachers from two counties in southeastern North Carolina. During 1997 and 1998, INSTRUCT was funded through the Eisenhower Professional Development Program to train 27 mathematics teachers from nine North Carolina counties. For the 1999-2000 school year, INSTRUCT will be supported by a U.S. Department of Education Technology Challenge grant to train up to 35 middle and high school mathematics teachers throughout North Carolina. Since 1997, the project has been granted licensure renewal credit by the state Mathematics and Science Education Network in recognition of the time participants spend online using INSTRUCT. Additionally, some school districts have granted technology renewal credit, which is part of the licensure renewal requirement for North Carolina teachers.

INSTRUCT TRAINING

The INSTRUCT model of training employs a two-week cycle for each of six standards; therefore, the training takes a total of 12 weeks, usually during the fall semester. Each cycle includes online reading assignments, an initial chat for asking questions and brainstorming implementation ideas, incorporation of these ideas in activities and/or lesson plans, execution of a standards-based lesson or activity in the classroom, and a final chat for reflecting on lessons learned from implementation. Participants close out their time with each of the standards by submitting an online Check for Understanding form. At the end of the semester, participants turn in to the UNCW training coordinator a portfolio of lesson plans and activities used during the semester, as well as samples of student work and an optional videotape. It is estimated that participants spend approximately 30 hours online and 12 hours offline completing training during a typical semester.

Following is a listing of the options available through INSTRUCT with a description of their function:

Check Here for the Latest News

This option gives the training coordinator a means of making announcements of online meetings and other events that do not necessarily require responses from participants.

A Hypermedia Version of Standards for Teaching Mathematics (NCTM, 1991)

This choice links users to a Web page with the following menu items: Worthwhile Mathematical Tasks, Teacher's Role in Discourse, Tools for Enhancing Discourse, Student's Role in Discourse, Learning Environment, and Analysis of Teaching and Learning. Each of these sub-menu items links to Web pages that employ multimedia to provide the user with an introduction to the NCTM *Standards for Teaching Mathematics* using authentic classroom activities and materials.

Each sub-menu page contains its own Check for Understanding form for users to complete and submit to the UNCW training coordinator in order to assess mastery of the standard. Assignment of licensure renewal credit is in part dependent upon successful responses to questions about the standards and on teachers' reports about their implementation of classroom activities intended to reflect the standards.

Online Educational Resources

Choosing this menu item takes the user to a table of links for education-related sites on the World Wide Web. Sites are grouped by category, such as Geometry and Chaos, History of Mathematics, Internet Project Ideas, Lesson Planning, National Agencies and Information Sources, Statistical Data Sources, and Technology Resources. Additionally, this page provides direct links to often-visited sites such as the Math Forum and K-12 Schools Online.

Attend a Meeting

This menu item leads the user to a chat area located on the UNCW WebBoard™. As indicated, participants are given assignments to carry out in their own mathematics classes in order to promote teacher active participation in and application of INSTRUCT training. An intentional by-product of these assignments is the encouragement of sustained interaction and collaboration among participants and facilitators (Honey & McMillan, 1994; Novick, 1996).

Teachers meet online in groups of ten or fewer, and the chats are facilitated either by a lead teacher who has already been through INSTRUCT training or by the UNCW training coordinator. The meeting option provides a means for teachers to brainstorm and to reflect together on implementing the standards in their classes, and for INSTRUCT facilitators to provide additional professional development materials and support to users. Participation in these meetings is another requirement for receiving licensure renewal credit. Access to the chatroom is limited to INSTRUCT participants and those non-participants who have been granted access by the UNCW training coordinator.

Join in a Discussion

This option leads participants to a threaded hypermail message board, also located on the UNCW WebBoard™, allowing them to communicate asynchronously with other INSTRUCT users at their own convenience. The format benefits teachers by involving them in more long-term discussions about issues raised in the NCTM standards, by facilitating the sharing of news and other items of interest between colleagues, and by affording users continual access to previous communications via discussion histories. Access to the discussion board is open to the public.

Send a Message

This choice provides a Web form for communicating privately with the UNCW training coordinator.

RESULTS OF TRAINING

Results from the first three years of the project have been encouraging. Upon conclusion of their training, participants are requested to complete an online survey asking them to reflect on their experience with INSTRUCT, especially in terms of changes that have occurred in their practices and/or beliefs. Results over the years have included changes in teacher attitudes toward traditional testing, including the belief that alternative assessment is a viable option in mathematics classes; teachers seeing themselves for the first time in the role of a facilitator while recognizing the active role students can take in their own learning; teachers promoting the belief that writing should be employed in the mathematics classroom; and teachers realizing the need to analyze the effectiveness of teaching methods and outcomes. Figure 1 illustrates how the combination of INSTRUCT's components and the repeated cycles of training, implementation, and reflection are intended to impact teacher practice and belief.

Results of an April 1999 follow-up survey of the 1997 and 1998 participants indicated that changes in practices and beliefs reported at the end of

FIGURE 1. The INSTRUCT model of Web-based professional development.

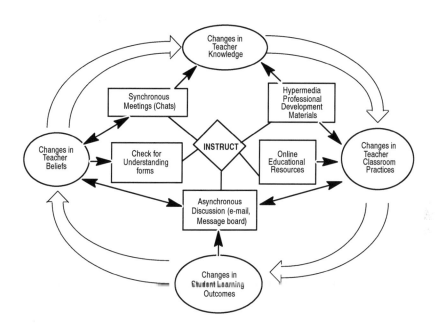

training seem to persist months and even over a year later. Table 1 displays those survey questions dealing with use of INSTRUCT training and resources in teaching mathematics, along with the number of responses for each of five categories (strongly agree, agree, disagree, strongly disagree, and no opinion). Of 15 respondents to the online survey (seven from 1997, eight from 1998), the majority of participants either agreed or strongly agreed with every statement. Perhaps the most striking result, in terms of the model in Figure 1, is that 60% of respondents agreed or strongly agreed that INSTRUCT training had a direct impact on their students' achievement. It is one thing for teacher practice to change for the duration of training, but it generally requires evidence of positive changes in student outcomes for changes in practice to persist and changes in belief to occur (Guskey, 1986).

Participants were also given the opportunity to share changes in practice that had taken place for each of the standards covered by the training. Of 90 possible responses (15 respondents, six standards each), specific examples of changes in practice were given in 64 cases, or 71% of the total possible responses. Following are some respondent examples of changes in practice:

TABLE 1. INSTRUCT Follow-Up Survey (n = 15)

	SA	A	D	SD	NO
1. I am actively using methods for enhancing discourse in my classes.	6	8	1	0	0
2. I actively seek to use alternative methods of student assessment in my classes.	4	8	3	0	0
3. I have students write on a regular basis in my classes.	4	6	5	0	0
4. I regularly access the online resources contained in INSTRUCT.	2	11	2	0	0
5. I believe I have become a resource for my department/school as a result of INSTRUCT training.	4	7	3	0	1
6. I have seen improvement in my students' achievement which I can directly attribute to changes I have made in my teaching as a result of INSTRUCT training.	2	7	2	0	4

"[I use] less paper and pencil analysis; more rubric-based grading of problem-solving tasks"; "I have used the graphic calculator and the Internet much more effectively in my classroom"; "[I now use] alternate methods of questioning and more open ended questions"; "I will now have students read the next lesson for homework, and they have to write comments, examples, and questions that they may have"; "[I use] more probing questions to have students explain their thinking"; "I am tolerating a much more loose and open structure to allow discourse to occur in a more relaxed, less rigid environment."

BENEFITS OF TRAINING

Results from the project highlight three specific ways in which mathematics teachers can benefit from Web-based professional development:

Consistent Opportunities for Reflection and Sharing with Colleagues

INSTRUCT appears to be capable of promoting the formation and maintenance of a virtual community, primarily through the use of synchronous chats. The ability to reflect on successes and failures with colleagues each week makes it more likely that implementation attempts will be sustained and not just one-time efforts. It also appears that the community-building aspect of Web-based professional development has at least some long-term effect beyond the actual training. In the follow-up survey mentioned earlier, 73% of respondents indicated they either agreed or strongly agreed with the statement, "I maintain regular contact with at least one fellow participant from the INSTRUCT training." It is notable that over half of those participants responding affirmatively to this item had completed their training more than a year prior to the survey.

Shortened Cycle for Training, Implementing and Evaluating New Practices

A key aspect of professional development on the Web is the natural capability for participating in training during the school year, either from school or from home. This gives teachers a realistic chance to plan classroom implementations that might be attempted the next day or the next week, instead of the typically longer time period between summer face-to-face training and classroom application. There is, however, a potential conflict of interest between allowing a shorter cycle of training and not having enough time to fully examine the impact of new practices. One of the participants summarized it this way: "What I enjoyed most about INSTRUCT was trying out the standards for two weeks and discussing with the different participants the things that they tried in their classrooms and the results that they achieved. I think that having more than two weeks to try a standard would have given me more time to test the effectiveness of the things I tried." Over the past three years a few participants have requested additional time to pursue particular standards in greater depth, and this kind of closer examination has always been encouraged.

Just-in-Time Support for Teacher Change

The availability of INSTRUCT's resources on the Web means that teachers have dependable access to information and colleagues who can assist with the implementation of new classroom practices. Although teachers may be hesitant to implement some aspect of a standard in their classes, primarily due to lack of familiarity, reviewing lesson plans and activities that others have used or listening to colleagues share their own experiences often makes the implementation process more feasible. As one teacher put it, "I loved the fact that I could go to INSTRUCT and find an activity or lesson on just about any topic. I also enjoyed the fact that I had a bond with several other teachers, and if I needed ideas, I could chat with one of them to get some different perspectives." The combination of immediate, personalized feedback and the Web's capability for providing up-to-date data and information makes Web-based professional development a powerful tool for changing teacher practice.

LIMITATIONS OF TRAINING

Some limitations to Web-based professional development have also been discovered during the time of the project. Successful Web-based instruction depends heavily on the presence of self-directed learning (Shotsberger, 1997). Since teachers essentially carry out an independent study of INSTRUCT's hypermedia materials, participants benefiting the most from the project tend to be self-starters who already possess a desire to improve their classroom practices. Similarly, participants need a base of experience from which to evaluate and discuss ideas for classroom implementation. This type of technology would be of limited utility for a new teacher who possesses neither the time nor the mature judgment to take full advantage of the variety of resources offered.

CONCLUSION

INSTRUCT employs the capabilities of the World Wide Web to provide practicing mathematics teachers with online professional development. INSTRUCT's features include a hypermedia version of the *Standards for Teaching Mathematics* (NCTM, 1991), links to online educational resources, and the capability to chat synchronously and to post messages asynchronously using WebBoard™. Three potential benefits associated with Web-based professional development have emerged from three years of experience with INSTRUCT: consistent opportunities for reflection and sharing; a shortened cycle for training, implementation, and evaluation; and just-in-time support for teacher change. As educational applications of Web-based professional development expand over the coming years, it appears likely that the vast potential of this new technology will dramatically change the landscape of mathematics teacher training.

REFERENCES

Guskey, T.R. (1986, May). Staff development and the process of teacher change. *Educational Researcher, 15*, 5-12.

Honey, M., & McMillan, K. (1994). Case studies of K-12 educators' use of the Internet: Exploring the relationship between metaphor and practice. *Machine-Mediated Learning, 4(2&3)*, 115-128.

National Council of Teachers of Mathematics (1991). *Professional standards for teaching mathematics*. Reston, VA: Author.

Novick, R. (1996). Actual schools, possible practices: New directions in professional development. *Educational Policy Analysis Archives, 4*(14). [Online]. Available: *http://olam.ed.asu.edu/epaa/v4n14.html*

Owston, R. (1998). *Making the link: Teacher professional development on the Internet*. Portsmouth, NH: Heinemann.

Romiszowski, A.J. (1997). Web-based distance learning and teaching: Revolutionary invention or reaction to necessity? In B. Khan (Ed.), *Web-based instruction* (pp. 25-37). Englewood Cliffs, NJ: Educational Technology Publications.

Shotsberger, P.G. (1997). Emerging roles for instructors and learners in the WBI classroom. In B. Khan (Ed.), *Web-based instruction* (pp. 101-106). Englewood Cliffs, NJ: Educational Technology Publications.

Lynda R. Wiest

The Role of Computers
in Mathematics Teaching
and Learning

SUMMARY. Computers can be powerful aids to mathematics teaching and learning. Changes brought about by the availability of these tools and the demands of an increasingly technological society impact curricular content and pedagogy in mathematics education as well as the very nature of mathematical thinking and understanding. This article presents ways in which technology is changing mathematics education, guidelines for appropriate technology use in the mathematics classroom, the impact of computers on mathematics learning, common uses of computers in mathematics education, and issues and concerns related to technology use in mathematics. *[Article copies available for a fee from The Haworth Document Delivery Service: 1-800-342-9678. E-mail address: <getinfo@haworthpressinc.com> Website: <http://www.HaworthPress.com>* © *2001 by The Haworth Press, Inc. All rights reserved.]*

KEYWORDS. Computers, technology, computer use, technology use, mathematics education, mathematics, teaching, learning

Computers play an increasingly significant role in mathematics teaching and learning, so much so that they are considered an important force behind the evolution of mathematics education. As Heid (1997) asserts, "The single most important catalyst for today's mathematics education reform movement

LYNDA R. WIEST is Professor, Department of Curriculum and Instruction, University of Nevada, Reno, NV 89557-0029 (E-mail: wiest@unr.edu).

[Haworth co-indexing entry note]: "The Role of Computers in Mathematics Teaching and Learning." Wiest, Lynda R. Co-published simultaneously in *Computers in the Schools* (The Haworth Press, Inc.) Vol. 17, No. 1/2, 2001, pp. 41-55; and: *Using Information Technology in Mathematics Education* (ed: D. James Tooke and Norma Henderson) The Haworth Press, Inc., 2001, pp. 41-55. Single or multiple copies of this article are available for a fee from The Haworth Document Delivery Service [1-800-342-9678, 9:00 a.m. - 5:00 p.m. (EST). E-mail address: getinfo@haworthpressinc.com].

41

is the continuing exponential growth in personal access to powerful computing technology" (p. 5). This article presents ways technology is changing mathematics education, guidelines for appropriate technology use in the mathematics classroom, the impact of computers on mathematics learning, common uses of computers in mathematics education, and issues and concerns related to technology use in mathematics.

CHANGES IN MATHEMATICS TEACHING AND LEARNING

Van de Walle (1998) outlines three ways technology is changing the nature of mathematics education. The first is that some mathematics skills have decreased in importance. Time taken to perform tedious paper-and-pencil computations, such as long division or constructions such as graphical representations, can be put to better use in more reasoning- and interpretation-oriented endeavors. This approach mirrors ways technology is used in everyday life. Second is the pedagogical idea that mathematics can be taught more effectively using computers. For example, visual and contextual representations that might not otherwise be available can be included. And teachers can use computer-based simulations to provide students with opportunities to work on problem situations that are difficult to experience without technology. As stated in *Principles and Standards for School Mathematics*, "Students can learn more mathematics more deeply with the appropriate use of technology" (National Council of Teachers of Mathematics, 2000, p. 25). Third, some mathematics topics and skills are more accessible or can receive greater emphasis. Data analysis is a prime example. The Internet provides access to a plethora of information that–combined with data-analysis tools and computer-generated graphs and tables–allows children to gather, represent, analyze, and interpret data at earlier ages and in expanded ways. Adults also have the opportunity to gain insight into new mathematical methods. A discussion recently took place on a mathematics educators' listserv in which one participant shared a new technique for simplifying a radical under a radical, which he had just learned while using the algebra software *Maple* by Waterloo Maple. Motivated to investigate, he learned that the method had been around for centuries but is no longer taught in school. This led him and a friend to explore how the method worked.

These changes impact mathematics content and curriculum, instructional methodologies (including assessment), learning styles, and the nature of mathematical thinking and understandings (Connell & Abramovich, 1999; Heid, 1997; National Council of Teachers of Mathematics, 1998; Schwartz & Beichner, 1999). The National Council of Teachers of Mathematics (NCTM) (1998) contends, "When a curriculum is implemented, time and emphasis must be given to the use of technology to teach mathematics concepts, skills, and applications in the ways they are encountered in an age of ever increasing access to more-powerful technology." Changes in favor of greater use of computers in mathematics education align nicely with other methodological emphases presently espoused by experts in the field, most notably, student responsibility for their own learning (Heid, 1997). Students become more autonomous, teachers more facilitative, and learning more authentic during carefully designed, computer-based projects. This student-centered environment also lends itself–contrary to popular belief about computer use–to collaborative group work. Students can pursue mathematics-oriented goals in dyads at the computer, or they can work independently and then share the results of their work with other students. Finally, technology can naturally support interdisciplinary learning by situating mathematics concepts in contexts and providing access to global data and communications. It is important to note that the classroom teacher's philosophy of mathematics education, and its implementation, ultimately shapes classroom learning.

APPROPRIATE USES OF TECHNOLOGY
IN THE MATHEMATICS CLASSROOM

Computer programs may be used for work with various mathematics concepts, including formulas, constructions, and proofs. Computers can also be used for accessing information and communicating with others mathematically. Whatever the uses of computers in mathematics, the focus should be on higher order thinking with an emphasis on inquiry, reasoning, and engagement in worthwhile mathematical tasks. Algebra instruction is one area that has been particularly transformed by use of Computer Algebra Systems (CASs), such as *Derive* (Texas Instruments), *Maple* (Waterloo Maple), *Mathcad* (MathSoft), and *Mathematica* (Wolfram Research). Students explore mathematical patterns and relationships while the computer performs most of the symbol manipulations and graphical representations. Algebraic thinking is not limited to the middle grades and above. Elementary students may also explore algebra concepts–such as functions–through age-appropriate software (see, for example, Sarama & Clements, 1998).

Used properly, technology fosters rather than replaces mathematical understandings (NCTM, 2000). Technology should be used as a medium that facilitates powerful mathematics learning rather than as an object of instruction in and of itself. It is a mechanism for accessing and processing information, modeling and exploring mathematics, and conducting mathematical investigations. In keeping with its role as a tool, teachers need to help students learn to decide when and how to use technology.

Using computers mainly for lower-level applications, such as computations, or for demonstrations unaccompanied by exploration and reasoning severely undercuts their potential to expand and challenge student thinking and may in fact limit it. Computers hold little advantage over their nonelectronic counterparts when they are used as the posers of problems and possessors of all answers to those problems (Cuoco & Goldenberg, 1996). Higher levels of thinking are attained when students pose "what if" questions before they test ideas, such as changing inputted information on a spreadsheet (e.g., a defined rule) or data in a graphed algebraic equation and observing the results. Followed by "why" questions, these types of analyses can develop more generalized thinking (Abramovich & Nabors, 1998). The NCTM (2000) discusses appropriate use of computers in mathematics education:

> The effective use of technology in the mathematics classroom depends on the teacher. Technology is not a panacea. As with any teaching tool, it can be used well or poorly. Teachers should use technology to enhance their students' learning opportunities by selecting or creating mathematical tasks that take advantage of what technology can do efficiently and well–graphing, visualizing, and computing. (pp. 25-26)

Connell's (1998) research findings from work with 52 rural elementary students point to the importance of a classroom matching technology use to its underlying instructional philosophy. For example, a student-centered, inquiry-oriented classroom that uses technology primarily as "a delivery system for prepackaged instruction" (p. 332) represents a misalignment of instructional practice that can negatively impact student learning. In short, using technology appropriately means emphasizing mathematical thinking in an environment with well-integrated instructional approaches grounded in current mathematics education theory.

COMPUTER IMPACT ON MATHEMATICS LEARNING

Computers have many features that can enhance student learning. Their multimedia capabilities lend a sensory component that may help reinforce concepts and appeal to a wider variety of learning styles. Graphical aspects help students visualize two- and three-dimensional geometric figures and represent mathematical ideas such as the nature of arithmetic versus exponential growth. Further, students can make conjectures and experiment with these graphical representations to *see* the results. In dynamic, interactive geometry programs, students may directly manipulate figures that remain intact as they change shape in continuous fashion, allowing students to see intermediate states. Figure 1 shows how a pentagon created on *Geometer's Sketchpad* (Key Curriculum Press) was reshaped on an adjacent copied version by dragging vertex G upward. Students can discover that the sum of the

FIGURE 1. Pentagon drawn on *Geometer's Sketchpad* with vertex G manipulated in copied version on right. Differences between diagonals on convex and concave polygons may be observed.

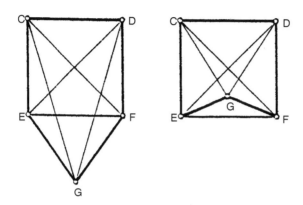

measures of the interior angles remains the same in such cases and explore differences between the diagonals of convex and concave polygons (the latter having at least one diagonal that falls outside the polygonal region). Successful mathematicians often possess an ability to visualize mathematical ideas. Computers extend this ability to a much broader audience, allowing students "opportunities to tinker with mathematical objects just as they might tinker with mechanical objects" (Cuoco & Goldenberg, 1996, p. 17). This use of computers fits nicely with the current emphasis on use of manipulative materials in mathematics education.

Testing one's thinking and seeing the results, as noted, can be a powerful way to bolster and stretch mathematical ideas. Students can generate a large set of instances from which they try to generalize, thus illustrating computing potential for pursuing the abstract and the general. Cuoco and Goldenberg (1996) issue the caveat, however, that students can have too much confidence in computer results and move on too quickly after making assumptions from a diverse set of cases. This, of course, points to the importance of the teacher's role in facilitating appropriate use of technology.

Another important computer feature conducive to teaching and learning is the potential for individualization. Good computer programs allow students to work at their own pace and give useful, immediate feedback on student performance. Numerous studies have found that personalizing problems–inserting individual students' names or other personal information into word problems–has a positive effect on student interest and achievement in solving the problems (e.g., López & Sullivan, 1991, 1992). Computers have been used in some of these studies to generate the personalized problems, demonstrating another way to individualize instruction.

Certainly, computers possess other valuable features that give them instructional appeal. Two of note are their interactive capabilities and their motivational qualities. Numerous studies show that computer use–conducted appropriately–has a positive impact on student achievement and other educational outcomes (e.g., attitudes) in mathematics education (Connell, 1998; Connors, 1995; Educational Testing Service, 1999). The Educational Testing Service (ETS) (1999) conducted a review of the results of the 1996 National Assessment of Educational Progress, which included national samples of 6,227 fourth graders and 7,146 eighth graders. The ETS reported that computer use improved both student proficiency in mathematics and the learning environment of the school, as measured by reduced student and teacher absenteeism and increased teacher morale. However, these benefits were found to depend upon *proper* use of technology, which had a greater effect than *frequency* of use. In particular, the report states that teachers need to focus computer use on applying higher order skills. Both this report and research conducted by DeVaney (1996) indicate that using computers frequently and for lower level skills (e.g., drill and practice) may be detrimental to students. The frequent-use factor, the ETS hypothesizes, might result from unproductive computer use, such as playing noneducational games.

COMMON USES OF COMPUTERS
IN MATHEMATICS EDUCATION

Educators use computers in a variety of ways to aid mathematics instruction. These uses can be broadly categorized into tool software, instructional software, the Internet, and programming.

Tool Software

Tool software is used as an aid toward another goal. It does not teach but rather performs a function that facilitates attainment of some objective. Examples of tool software are spreadsheets (e.g., *Microsoft Excel* or Davidson and Associate's *The Cruncher*), data analysis and representation software (e.g., Sunburst's *Data Explorer* or Tom Snyder's *The Graph Club*), and dynamic geometry software (e.g., Key Curriculum Press's *The Geometer's Sketchpad*). Word processing is a less mathematical but nevertheless viable type of tool software for use in mathematics education. Word processors can be used to write mathematics journals in which students make entries such as explaining and justifying problem solutions, relating their understanding of a concept, or preparing a mathematics project report. Students may also write their own mathematics problems, an important exercise in conceptual, higher-level thinking. Some tool software is general-purpose and other is subject-specific. Connell and Abramovich (1999) suggest broader and more sophisti-cated uses of general-purpose software, such as spreadsheets, as one response to the economic constraints associated with technology-access issues. One Web site that offers secondary activities using tool technology may be found at *http://curry.edschool.virginia.edu/teacherlink/math.* The site was developed at the University of Virginia as a result of a funded project.

Dugdale (1998) presents one effective spreadsheet activity for use with middle-grades through precalculus students. Students explore common num-ber sequences, allowing them to investigate patterns and develop number sense. They then place the sequences in separate columns on a spreadsheet and create formulas for each, with the advantage of being able to generate and test ideas while receiving immediate feedback. The number sequences are shown in horizontal lists below and then in Dugdale's display represent-ing the formulaic spreadsheet version (Figure 2).

Counting numbers: 1, 2, 3, 4, 5, 6, 7, 8, 9, 10, 11, 12, 13 . . .
Square numbers: 1, 4, 9, 16, 25, 36, 49, 64, 81, 100, 121, 144, 169 . . .
Triangular numbers: 1, 3, 6, 10, 15, 21, 28, 36, 45, 55, 66, 78, 91 . . .
Fibonacci numbers: 1, 1, 2, 3, 5, 8, 13, 21, 34, 55, 89, 144, 233 . . .

Instructional Software

Instructional software is designed to teach students skills and concepts. It is loosely classified and discussed below in four overlapping categories: drill and practice; problem solving, simulations, and games; concept instruction; assessment, remediation, and tutorial. General guidelines for choosing good software can be found in the October 1999 issue of *Technology and Learning* (Branzburg & McLester, 1999), and an instrument for evaluating educational software is available online from the Ohio SchoolNet (*http://www.enc.org/rff/ ssrp/docs/ssrpinst.htm*).

Drill and practice. Drill-and-practice software does what the name im-plies–provides practice on previously taught skills and concepts, such as giving answers to basic facts (those with single-digit addends or factors and their subtraction and division inverses) or identifying fraction equivalents. It is the most common type of software available for elementary mathematics (Van de Walle, 1998). Drill-and-practice software does not represent one of the more powerful uses of computers for mathematics instruction. Used properly, these programs can have value, but they "are not the unique contribution of technology and do not represent a genuinely new opportunity" (Goldenberg, 1998, Online).

Problem solving, simulations, and games. These types of software packages emphasize attaining goals in reasoning-oriented, decision-making contexts, which may be real-world settings. Problem-solving programs can involve word problems or more involved pursuits. Simulations, true to their use in science, can parallel or approximate experiments with real systems or phenomena that cannot be done or are impractical to do in actuality. For example, students might participate in a computer simulation of the stock market, in which they "buy" and "sell" stocks without investing real money. Games may be drill-and-practice or they may require higher order thinking. They often include a competitive element and can be quite motivational by design.

Concept instruction. Some programs are designed to teach specific mathematics concepts, such as ideas about fractions, probability, or algebra. This software type is intended to mimic regular classroom instruction in teaching

FIGURE 2. Formulaic spreadsheet version of common number sequences (Dugdale, 1998, p. 206). Arrows indicate that formulas extend downward to continue the numeric patterns.

	A	B	C	D
	NUMBER SEQUENCES			
1				
2	Counting Numbers	Square Numbers	Triangular Numbers	Fibonacci Numbers
3	1	= A3 * A3	1	1
4	= A3 + 1		= C3 + A4	1
5				= D3 + D4
6				
7				
8				
9				
10				
11				
12				
13				
14	▼	▼	▼	▼
15				

material as if it were new to the learner. Therefore, it is likely to incorporate currently popular instructional methods, such as use of mathematics manipulatives. Programs in this category may consist of sequential lessons that constitute a curriculum unit. Tutorial programs, listed in the next category, might very well be included here; however, they are separated due to their greater orientation to individualized instruction, much in the way that personal tutoring is distinguished from whole-class instruction. Tutorials may be more likely to involve independent, self-paced learning that adapts itself to and provides more feedback about student performance. Concepts addressed may or may not be assumed to be new to the learner.

Assessment, remediation, and tutorial. Some computer programs are designed to assess student performance. These types may be used to learn more about student thinking, for example, by providing insight into the quality of students' geometry and trigonometry knowledge schemas (cf. Chinnappan, Lawson, & Gardner, 1998). This information may or may not be used as formative evaluation that leads to remediation. Other software is designed to diagnose and remediate. Parvate, Anjaneyulu, and Rajan (1998) developed the Mathemagic system to diagnose individual performance and provide customized remediation for high school students. Their research on the program's use found substantial improvement for the weaker students but little effect upon above-average students. The relatively short amount of time in which these effects occurred led the authors to conclude that using computers for remedial teaching can provide students with relatively low-cost, efficient support. Finally, tutorial programs teach concepts through use of a general set of problems for everyone or through remedial work that adapts instruction to user input and needs. Tutorial programs may maintain records of student performance. Note that a good drill program can be tutorial.

The Internet

The Internet can be used in many important ways in mathematics instruction. It is a seemingly limitless and continually updated (and outdated) source of information that can provide data for mathematical investigations. It also provides access to the types of experiences described above under instructional software. Innumerable good sites exist for teachers to get lesson plans, share ideas with other teachers, and hone their own mathematics skills. Teachers may also maintain connections to professional organizations, such as the National Council of Teachers of Mathematics (*http://www.nctm.org*). Students may apply, remediate, and extend their mathematics knowledge, access data for school tasks, and communicate with subject-matter specialists, online mentors, or other students. Various sites exist, for example, for students to get homework help. Probably the most highly recommended mathematics Web site, in general, is The Math Forum (*http://forum.swarthmore. edu*), based at Swarthmore College, Pennsylvania.

Communication via the Internet can take place by participating in mathematics-oriented chat rooms on the Web or e-mail listservs, or by establishing key pals–the modern version of pen pals–from around the globe. Collaborative projects may be conducted with students from one's own or other countries. One teacher conducted a project in which his students measured the shadow of a two-meter pole at different times of the year in conjunction with their study of seasonal changes (Classroom Connect, 1996). Several other classes from different latitudes around the world participated by reporting their own data regularly to the teacher, who collated the data and sent it to partner classrooms so they could chart the change over time in various latitudes. (See also Lynes, 1997, regarding Internet mathematics projects.) A good site for teachers interested in online collaborative projects and expeditions is Lightspan.com (*http://www.lightspan.com*). (Consult its "Site Guide for Teachers.")

The NCTM (1998) says that every mathematics classroom should have computers with Internet connections. Some general benefits that may accrue from Internet use are infusion of a global/multicultural perspective into mathematics learning and appreciation for the relevant, real-world nature of mathematics. Results from Gerber and Shuell's (1998) research on students' Internet use points to the importance of helping students become familiar with the Internet and of teaching them to conduct effective searches for information before beginning a project. Further, teachers should be sure students have the prerequisite mathematics knowledge and skills for the project and should structure discussion and reflection in ways tailored to the project's instructional objectives, such as an appreciation for real-world uses of mathematics.

Programming

Computer programming may be valuable for students for several reasons (Cuoco & Goldenberg, 1996; Van de Walle, 1998). Students must understand the mathematics involved. They also must exercise logical, sequential thinking, with associated planning skills, and learn the importance of language precision. Logo is a popular computer language that allows students of all ages to practice programming in an interesting way. One geometry-related use is to write commands to move a "turtle" (which sometimes appears as a tiny triangle) around the screen to create shapes and designs. For example, FD 30 moves the turtle forward 30 units (about 3 cm), drawing a straight line along the traveled path. RT 40 commands the turtle to turn 40 degrees to the right. A REPEAT function allows for doing a series of identical actions. For example, REPEAT 4 [FD 60 RT 90] creates a square with 60-unit sides. Mathematics knowledge and skills that may be exercised in these types of activities include angle measures and relative lengths, properties of shapes, spatial abilities, and patterning. One elementary teacher incorporated a multi-cultural perspective into programming when she had her students make a Navajo blanket design and then write a Logo program to construct the pattern (Bradley, 1993; see Figure 3). More research is needed in this area to determine the impact of programming on students' mathematical thinking and its transferability to other areas of mathematics (Heid, 1997).

ISSUES AND CONCERNS

A number of issues and concerns surround integration of technology into mathematics instruction (Heid, 1997). One is concern about how technology is used in the classroom, including what students might *not* learn (e.g., computational expertise). This relates to another issue–that of public perception. Public fear of misuse of technology has the potential to "sideline" this promising instructional tool. These concerns might be addressed by striving to make appropriate use of technology and conducting effective public-relations efforts. Two other issues that bear greater discussion will be addressed briefly as they relate to technology use in mathematics education: educational equity and teacher preparation.

Educational Equity

The National Council of Teachers of Mathematics' (2000) equity principle holds that "excellence in mathematics education requires equity–high expectations and strong support for all students" (p. 12). Technology has the potential to help or hinder in this regard. On the positive side, it can support

FIGURE 3. Intermediate step in a Navajo blanket design drawn using Logo (Bradley, 1993, p. 521).

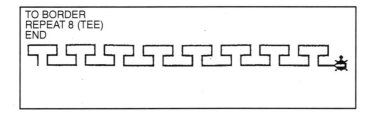

lower level technical skills (e.g., computation and graph construction) while providing access to higher level explorations, it can serve diagnostic and remedial functions, and it can connect students from various backgrounds with the world at large. It uses an instructional mode that may "reach" certain types of students, such as those who are highly distractible or have organizational difficulties, and it can include adaptive equipment helpful to students with physical disabilities. One interesting project aimed at grades four through seven created online lessons and activities on mathematics and aeronautics (Kraus, 1998). The project allowed both general and special education students to gain information about and interest in aeronautics-related careers. The project was founded on two premises concerning students with physical disabilities: awareness that these students may be at a disadvantage for using mathematics manipulative materials, on which strong emphasis is placed in today's classrooms, and that they might not otherwise consider or be prepared for possible careers in aeronautics.

On the "down" side, concern is continually voiced that economic constraints prevent some schools and homes from access to the benefits of modern technology, so much so that a widening gap might take place among schools or students of differing financial backgrounds. Findings from the 1996 National Assessment of Educational Progress (Educational Testing Service, 1999) show that Black, poor, urban, and rural students are less likely to be exposed to higher order uses of technology or to have a home computer. These students also tend to have mathematics teachers who are less prepared in technology use. Of course, Internet connections are also highly dependent on adequate finances.

Females' disengagement from and underrepresentation in technological courses and careers and their lack of experience using computers in powerful mathematical ways have been well documented (e.g., American Association of University Women, 1998; Makrakis & Sawada, 1996; Morahan-Martin, Olinsky, & Schumacher, 1992). In fact, reports of a major new gender gap were widely publicized in the fall of 1998 based on a research report released by the American Association of University Women (1998). Suggestions for engaging females in technology and guidelines for instructional techniques suitable for females using computers may be found in Ettenheim, Furger, Siegman, and McLester (2000). Among these are encouraging self-guided exploration, providing role models and mentors, establishing computer clubs for girls, and using the computer in a social-educational manner and toward worthwhile ends. Research shows that structured experience with technology can be particularly beneficial to females and members of other groups underrepresented in mathematics and science (Connors, 1995; Sherman & Weber, 1999). Concerted, continued work toward technological equity must remain an important goal in mathematics education.

Teacher Preparation

A vitally important concern often voiced in relation to the increasing use of technology in teaching is teacher preparation. Teachers need substantial, ongoing preparation in technology use with sufficient time to explore and convenient access to computers in order to gain the confidence and competence necessary to teach students to use it effectively. The Educational Testing Service (1997) says:

> School leaders report that the learning curve is steeper for teachers than it is for the children, and many have told us that the biggest mistake they made when introducing computers and other technologies into their classrooms was underestimating the amount of training the teachers would need. (p. 8)

Professional development of teachers in technology has been linked to student achievement and some aspects of school climate (Educational Testing Service, 1999). Therefore, strong efforts must be made to ensure continuous, quality teacher education in use of technology, at both pre-service and in-service levels. Members of the Association of Mathematics Teacher Education listserv recently debated the need for specific "technology in mathematics education" courses in teacher education programs. Proponents noted that such technology use might not be sufficiently addressed otherwise (e.g., in methods classes), due mainly to time constraints and insufficient specialized knowledge on the part of the instructor. Goldenberg (1998) notes: "Computers are nothing but what they're used for, and what they're used for changes constantly . . . The teachers . . . remain constant novices" (p. 1). The Educational Testing Service (1999) recommends that schools reserve one-third of their technology budget for teacher preparation.

CLOSING COMMENTS

Computers are changing the face of mathematics education, if not mathematics itself. Great effort will be required to ensure that mathematics education shadows and anticipates the way technology is changing the present and–especially–will change the future. Technology should be as integral to mathematics education as it is to the world beyond the classroom, rather than amounting to a curricular add-on. A statement I recently observed posted on a colleague's door sums up this charge: "We educate students for their future, not our past."

REFERENCES

Abramovich, S., & Nabors, W. (1998). Enactive approach to word problems in a computer environment enhances mathematical learning for teachers. *Journal of Computers in Mathematics and Science Teaching, 17,* 161-180.

American Association of University Women. (1998). *Gender gaps: Where schools still fail our children.* Washington, DC: Author.

Bradley, C. (1993). Teaching mathematics with technology: Making a Navajo blanket design with Logo. *Arithmetic Teacher, 40,* 520-523.

Branzburg, J., & McLester, S. (1999). Advice for picking out great software. *Technology and Learning, 20*(3), 44.

Chinnappan, M., Lawson, M. J., & Gardner, D. (1998). The use of microcomputers in the analysis of mathematical knowledge schemas. *International Journal of Mathematical Education in Science and Technology, 29,* 805-811.

Classroom Connect. (1996). *Internet curriculum integration: Creating Internet projects* [video series tape 3]. Lancaster, PA: Author.

Connell, M. L. (1998). Technology in constructivist mathematics classrooms. *Journal of Computers in Mathematics and Science Teaching, 17,* 311-338.

Connell, M. L., & Abramovich, S. (1999, February-March). *New tools for new thoughts: Effects of changing the "tools-to-think-with" on the elementary mathematics methods course* [CD-ROM]. Paper presented at the Society for Information Technology and Teacher Education International Conference, San Antonio, TX. Available: SilverPlatter File: ERIC Item: 432 267

Connors, M. A. (1995). Achievement and gender in computer-integrated calculus. *Journal of Women and Minorities in Science and Engineering, 2,* 113-121.

Cuoco, A. A., & Goldenberg, E. P. (1996). A role for technology in mathematics education. *Journal of Education, 178*(2), 15-32.

DeVaney, T. A. (1996, November). *The effects of computers and calculators on computation and geometry achievement* [CD-ROM]. Paper presented at the Annual Meeting of the Mid-South Educational Research Association, Tuscaloosa, AL. Available: SilverPlatter File: ERIC Item: 404 163

Dugdale, S. (1998). A spreadsheet investigation of sequences and series for middle grades through precalculus. *Journal of Computers in Mathematics and Science Teaching, 17,* 203-222.

Educational Testing Service. (1997). *Ten lessons every educator should know about technology in the classroom.* Princeton, NJ: Author. (Retrieved April 12, 2000, from the World Wide Web: *http://www.ets.org/download.html*)

Educational Testing Service. (1999). *Does it compute? The relationship between educational technology and student achievement in mathematics.* Princeton, NJ: Author. (Retrieved April 12, 2000, from the World Wide Web: *http://www.ets.org/research/pic/dic/techtoc.html*)

Ettenheim, S. G., Furger, R., Siegman, L., & McLester, S. (2000). Tips for getting girls involved. *Technology and Learning, 20*(8), 34-36.

Gerber, S., & Shuell, T. J. (1998). Using the Internet to learn mathematics. *Journal of Computers in Mathematics and Science Teaching, 17,* 113-132.

Goldenberg, E. P. (1998). Chipping away at mathematics: A long-time technophile's worries about computers and calculators in the classroom. (Retrieved April 12, 2000, from the World Wide Web: *http://forum.swarthmore.edu/technology/papers/papers/goldenberg/goldenberg.html*)

Heid, M. K. (1997). The technological revolution and the reform of school mathematics. *American Journal of Education, 106*, 5-51.

Kraus, L. E. (1998, March). *Teaching mathematics to students with physical disabilities using the World Wide Web: The Planemath Program* [CD-ROM]. Paper presented at the California State University-Northridge Conference, Los Angeles, CA. Available: SilverPlatter File: ERIC Item: 421 815

López, C. L., & Sullivan, H. J. (1991). Effects of personalized math instruction for Hispanic students. *Contemporary Educational Psychology, 16*, 95-100.

López, C. L., & Sullivan, H. J. (1992). Effect of personalization of instructional context on the achievement and attitudes of Hispanic students. *Educational Technology Research and Development, 40*(4), 5-13.

Lynes, K. (1997). Mining mathematics through the Internet! *Teaching Children Mathematics, 3*, 394-396.

Makrakis, V., & Sawada, T. (1996). Gender, computers and other school subjects among Japanese and Swedish students. *Computers and Education, 26*, 225-231.

Morahan-Martin, J., Olinsky, A., & Schumacher, P. (1992). Gender differences in computer experience, skills, and attitudes among incoming college students. *Collegiate Microcomputer, 10*, 1-8.

National Council of Teachers of Mathematics. (1998). *The use of technology in the learning and teaching of mathematics* [position statement]. (Retrieved April 10, 2000, from the World Wide Web: *http://www.nctm.org/about/general.information/Position.Statement.13.htm*)

National Council of Teachers of Mathematics. (2000). *Principles and standards for school mathematics*. Reston, VA: Author.

Parvate, V., Anjaneyulu, K. S. R., & Rajan, P. (1998). *Mathemagic*: An adaptive remediation system for mathematics. *Journal of Computers in Mathematics and Science Teaching, 17*, 265-284.

Sarama, J., & Clements, D. H. (1998). Using computers for algebraic thinking. *Teaching Children Mathematics, 5*, 186-190.

Schwartz, J. E., & Beichner, R. J. (1999). *Essentials of educational technology*. Boston: Allyn & Bacon.

Sherman, S., & Weber, R. (1999). Using technology to strengthen mathematics and science instruction in elementary and middle schools. *Journal of Women and Minorities in Science and Engineering, 5*(1), 67-78.

Van de Walle, J. A. (1998). *Elementary and middle school mathematics: Teaching developmentally* (3rd ed.). New York: Longman.

Robert Mayes

CAS Applied in a Functional
Perspective College Algebra Curriculum

SUMMARY. *Act in Algebra* (ACT) is a functional perspective curriculum for college algebra that focuses on the use of realistic applications to motivate the acquisition of mathematical concepts through active student explorations. A computer algebra system (CAS) is used as an open-ended tool in the exploration of mathematics and the modeling of data. The goal of ACT is to improve student cognition, metacognition, and affect with respect to college algebra. A series of studies was conducted from the spring of 1994 to the spring of 1999 to determine the effects of the CAS in a functional perspective curriculum. This article provides an overview of the findings of those studies and discusses the implications of these findings for the teaching and learning of algebra. *[Article copies available for a fee from The Haworth Document Delivery Service: 1-800-342-9678. E-mail address: <getinfo@haworthpressinc. com> Website: <http://www.HaworthPress.com> © 2001 by The Haworth Press, Inc. All rights reserved.]*

KEYWORDS. Algebra, problem-solving, computer algebra system, conceptual understanding, cognition, metacognition, affect, function concept

An important goal for teaching mathematics is the development of mathematical power in students–power that will enable students to explore, conjecture, reason, formulate and solve problems, and communicate mathematically

ROBERT MAYES is Associate Professor, College of Arts and Sciences, Department of Mathematical Sciences, University of Northern Colorado, Greeley, CO 80639 (E-mail: rmayes@bentley.unco.edu).

[Haworth co-indexing entry note]: "CAS Applied in a Functional Perspective College Algebra Curriculum." Mayes, Robert. Co-published simultaneously in *Computers in the Schools* (The Haworth Press, Inc.) Vol. 17, No. 1/2, 2001, pp. 57-75; and: *Using Information Technology in Mathematics Education* (ed: D. James Tooke and Norma Henderson) The Haworth Press, Inc., 2001, pp. 57-75. Single or multiple copies of this article are available for a fee from The Haworth Document Delivery Service [1-800-342-9678, 9:00 a.m. - 5:00 p.m. (EST). E-mail address: getinfo@haworthpressinc.com].

57

(Heid, 1995). Laudable goals, but what is the process for getting there? How can we empower college algebra students who have already failed to be empowered by their high school experiences? What is the role of a computer algebra system in that empowerment?

Bednarz, Kieran, and Lee (1996) identify four perspectives for teaching algebra that have come to the fore in reform curriculum around the world, each with the potential to empower students:

1. *Generalization Perspective:* generalizing patterns, both numeric and geometric, and the laws governing numerical relations
2. *Problem-Solving Perspective:* solving specific problems or classes of problems
3. *Modeling Perspective:* modeling of physical phenomena
4. *Functional Perspective:* focusing on the concepts of variable and function.

The generalization perspective has a strong basis in the historical development of algebra. It is hypothesized that both the Babylonian and Greek cultures developed algebraic concepts by generalizing arithmetic and geometric solution procedures. Modern-day students of mathematics also attempt to generalize arithmetic to understand algebra. Mason (1996) believes that "the heart of teaching mathematics is the awakening of pupil sensitivity to the nature of mathematical generalization and . . . to specialization." Radford (1996) expresses concern over the complex concept of validating student generalizations and the limiting perspective of generalization as a proof-process intrinsically bound to variable and formula. He contends that a problem-solving approach requires the fundamentally different concepts of unknown and equation. We will not pursue the generalization perspective due to its limitations in introducing a full range of algebraic concepts on the collegiate level, although we do not discount its importance as a bridge from arithmetic to prealgebra concepts.

The problem-solving perspective has its lineage in Viete's and Descartes' conversion of algebra from arithmetic synthesis, solving simple equations by reversing arithmetic operations, to algebraic analysis, the assumption that a problem is solved which allows manipulation of an unknown. This transformed algebra from Diophantus' search for solutions to specific problems to solving classes of problems with similar structure. The distinguishing characteristics of the problem-solving perspective are the existence of a problem, pure or applied, for which the student has no immediate algorithm and the forming and solving of equations representing that problem (Bell, 1996). Motivation for the student arises from the power of the algebraic method.

In contrast, both the modeling perspective and the functional perspective focus on the construction and interpretation of mathematical models. In the modeling perspective the problem may be real-world or based in abstraction;

the focus is on student construction of mathematical narratives that fuse aspects of events and situations with properties of symbols and notations (Nemirovsky, 1996). Students construct meaning for graphic, tabular, and symbolic representations and use them to interpret or describe phenomena. The problem situation in the functional perspective is often a real-world problem consisting of discrete data that are to be modeled by a function. This process serves as a means of introducing families of functions and the study of their properties (Heid, 1996).

Wheeler (1996) surmises that the generalization and modeling perspectives lack sufficient development to be the basis for a workable program. His concern with the generalization perspective is that it does not allow for a fully functioning symbolic system. He believes the only way that the modeling approach can be implemented with novice students is to trivialize the difficult concept of modeling. Wheeler supports the problem-solving and functional perspectives. In support of the problem-solving perspective, he states: "It has the sanction of history and a century of teaching tradition behind it" (p. 317). He notes that this approach directly supports what many people regard as the main purpose of mathematics: the solving of well-formulated problems. He concedes that, with the advent of the computer, the functional approach may be viable, but there is a need for research to support this contention.

ACT (*A*pplications, *C*oncepts, and *T*echnology) *in Algebra* (Mayes & Lesser, 1998) is a functional perspective curriculum utilizing the CAS Derive as a tool to actively engage students in modeling real-world problems. The qualitative and quantitative studies conducted on the use of *ACT in Algebra* were designed to determine the effects of this curriculum on students' cognition, metacognition, and affect with respect to learning algebra.

BACKGROUND AND THEORETICAL FRAMEWORK

Functional Perspective

The concepts of variable and function are central to the functional perspective in algebra. Fey and Heid (1995) developed a functional perspective curriculum for high school that epitomizes this approach:

1. Variables are used to describe real-world quantities that vary.
2. Functions are used to model the relationship between variables.
3. Families of functions, such as linear, quadratic, and exponential families, are studied as reasonable models of real-world phenomena.
4. Properties of functions are studied as they relate to interpreting real-world situations.

5. Conceptual understanding includes reasoning within and among multiple representations of function, including graphic, numeric, and symbolic representations.

Mayes and Lesser (1998) developed the functional perspective curriculum (FPC) ACT in Algebra for college algebra with similar goals.

The FPC introduces the student to the concept of function before equation. Functions are explored using graphic, numeric, and symbolic representations and are interpreted as a process and an object. Function models for discrete real-world data are constructed using finite differences, curve of best fit, and linearization techniques. Equations are presented as specific cases of function models, cases where the dependent variable is given. Topics included in the FPC are function, modeling, algebraic functions (polynomial, rational, radical, absolute value), transcendental functions (logarithmic and exponential), solving equations and inequalities, systems of equations and matrices, and the theory of polynomial equations (studied as properties of the family of polynomial functions). The FPC uses the CAS Derive as a tool to explore modeling and function concepts via 10 computer laboratories. Three of the labs require a written report with two submissions.

Nemirovsky (1996) questions the use and nature of real-world problems in the functional perspective. He identifies four expectations commonly held for the contribution of real-world problems to the learning of algebra:

1. Real-world problems enable students to approach new mathematical concepts using ideas and situations that are familiar to them.
2. Real-world problems suggest to students that the formal definitions adopted in algebra are natural and reasonable.
3. Real-world problems offer fruitful contexts for students' learning to deal with complexity.
4. Real-world problems help students to realize that algebra is useful and relates to actual issues of social, financial, or political relevance.

Nemirovsky dismisses the first expectation by questioning if real-world problems are "real" for a student. He contends that what is real for students is based upon their experiences, not on the content of the problem. He then argues that the second expectation is fulfilled by any problem that evokes a shared language with which discussions of subtleties of a concept occur. This shared language is evoked when a student actively engages in the solution of a problem, abstract or real. Finally, Nemirovsky dismisses the third expectation by arguing that complexity is not an exogenous factor defined by the problem, but emerges from the qualities that surround students' experiences.

Nemirovsky's arguments center on the individual student's perception of a real-world problem. Is the problem real, complex, and of intrinsic interest to the student? While we agree with concerns over student perception, our experience has been that the majority of college algebra students are not engaged by problems based in pure mathematics. Neither are they engaged by decontextualized problems posing as applications. In fact, it may well be that the fourth expectation is central to engaging students in the study of algebra. The next two sections establish a theoretical basis for the affective and cognitive issues addressed in this report.

Affective Issues

Research on mathematical problem-solving was heavily influenced by the theories and methods of cognitive science during the 1970s and 1980s (McLeod, 1994). The emergence of affect as an important part of cognition in mathematics education is due to Mandler (1989). McLeod (1992) summarizes Mandler's view:

> First, students hold certain beliefs about mathematics and about themselves that play an important role in the development of their affective responses to mathematical situations. Second, since interruptions and blockages are an inevitable part of the learning of mathematics, students will experience both positive and negative emotion as they learn mathematics; these emotions are likely to be more noticeable when the tasks are novel. Third, students will develop positive or negative attitudes toward mathematics as they encounter the same or similar mathematical situations repeatedly.

This developmental psychology view of affect parallels the constructivist view on cognition; that is, students experience emotions that develop into attitudes, which are then used to construct beliefs.

In reform curricula, novel tasks are used to promote critical thinking by the student. If critical thinking is to occur, the student must experience blockage; that is, the student cannot apply a known algorithm to solve the problem directly. The blockage will elicit an emotion from the student. Mandler hypothesizes that the repeated experience of positive or negative emotions will instill themselves as attitudes. Hart (1989) defines an attitude as a predisposition to respond in a particular way to a given stimulus. Attitudes are moderate in duration, intensity, and stability. Attitudes develop into beliefs over a long period of time. Beliefs are long term in duration, low in intensity, and relatively stable. McLeod (1992) states that beliefs are largely cognitive in nature. Beliefs about self and ability in mathematics, in turn, affect emotions.

According to Middleton (1995), personal interest, degree of personal control, and degree of arousal make mathematics motivating for a student. Also, "increasing a student's belief that mathematics is useful will often increase motivation and thus achievement" (Stage, 1992, p. 109). The curriculum implemented in these studies has a strong emphasis on modeling real-world situations in order to increase awareness about the usefulness of mathematics. It is important to note that, although students may view mathematics as useful, for many this is not enough to motivate them to learn mathematics. The main reason for this phenomenon may be a student's confidence in his or her ability to do mathematics.

It is very difficult to motivate a student who believes he or she does not have the ability to succeed in mathematics. Students tend to believe that mathematics learning depends more on ability than effort, and adults are willing to accept poor performance in mathematics more than in other subjects (McLeod, 1992). This belief would in turn influence the attitudes the students bring to the classroom and the emotions experienced during problem-solving.

The view students hold about mathematics has an important effect on their learning of mathematics. Most students believe that mathematics problems should be solved very rapidly; therefore, they tend to give up on problems that require more time than expected. Also, the student's view of mathematics as manipulation and memorization results in a search for an arithmetic operation, rather than trying to make sense of the problem (Schoenfeld, 1985). Such beliefs generate frustration when the student does not have an algorithm that provides the answer. The student fails to realize that frustration is a part of learning. So he or she is less likely to be active and persistent in the learning of mathematics.

Cognitive Issues

The function concept is perhaps the most important concept in all of mathematics and serves as the central theme in the functional approach. Yet the concept of function is remarkably complex, and comprehension of it presents a significant challenge for students. Among the numerous misconceptions of function held by students are:

1. The correspondence representing a function is a rule that is algorithmic.
2. The independent variable must change as the dependent variable is varied, so constant functions are not functions.
3. A function is a formula, not a graph or table.
4. A function is a 1-1 correspondence.
5. A function's graph cannot have corners (differentiability criteria) or holes (continuity criteria).
6. A function need not be univariant (have a unique image for each domain value).
7. Graphs are interpreted primarily as objects, while tables and formulas are interpreted as processes.

Vinner and Dreyfus (1989) suggest that misconceptions such as these are due to an unavoidable conflict between formal mathematical structure and the construction of concepts. Students may invoke a formal concept definition when pressed by the teacher, but they will elicit their own concept image (all of the facts, relationships, examples, and images they associate with the concept) when solving a problem. We will explore the effect of the functional perspective on a student's concept image.

Sfard (1995) argues that resistance to upheavals in tacit epistemological and ontological assumptions obstructed the progress of mathematics throughout history, and such obstructions are mirrored in the classroom. Historically, algebra progressed through three stages: rhetorical/syncopated, symbolic, and abstract algebra. She claims the rhetoric and syncopated expressions imposed an operational mode of thinking; that is, a process concept of function. The substantial burden placed on working memory was not relieved until modern symbolism was introduced. This symbolism was more efficient than rhetorical algebra, inducing a structural approach (object conception). A number of studies found that students revert to the rhetorical method despite years of symbolic algebra (Sfard, 1995), which supports the claim that the transition from an operational (process) to a structural (object) interpretation of function is inherently difficult. The transition is labeled *reification* or *encapsulation*: the creation of a mental object from a process or procedure. Sfard argues that new concepts should not be introduced using structural descriptions, which implies that introducing the function as a structural model of real-world problems could be cause for concern. We will study the process-object duality held by students in a functional perspective.

Cognitive science supports the view of understanding as the building of networks between related concepts. Thus comprehension of the function concept involves translation across the three primary representation systems: graphic, symbolic, and numeric (Slavit, 1994). Moschkovich et al. (1993) combined representations with interpretations (process-object) in an integrated view of function concept. We will determine the effect of the functional perspective on a student's abilities to translate between the three representations and the two interpretations of function.

Lastly, O'Callaghan (1998) uses the ability of a student to model a real-world problem, and the reverse procedure of interpreting functions in terms of real-world problems, as measures of students' understanding of the function concept. He found that a functional perspective resulted in improved understanding of the function concept, as well as modeling, interpreting, and translating. However, reifying was not significantly improved. We included modeling-interpreting of real-world problems as a measure of function understanding in this study. Thus several related theoretical perspectives measured the function concept: concept image, process-object duality, translation between representations, and the modeling-interpretation of real-world problems.

Research Questions

The purpose of these studies was to examine the effect of a functional perspective using a computer algebra system (CAS) on a student's affect and cognition in a college algebra course. The specific research questions related to this purpose are:

1. Will the functional perspective curriculum (FPC) affect students' affective attitudes and beliefs about mathematics?
2. Will the FPC affect students' cognitive concept of function with respect to concept image, process-object duality, translation between representations, and the modeling-interpretation of real-world problems?
3. Will the FPC affect students' metacognition with respect to problem-solving and real-world modeling?

METHODOLOGY

Subjects

The accessible population for the studies conducted from 1994 to 1999 consisted of all college algebra classes at a mid-sized university in the Rocky Mountain region. Total enrollment averaged 300 students per semester. Early studies on the function concept and affect were primarily quantitative in nature, collecting data from all students in the accessible population. A measure of students' understanding of the function concept and a survey of attitudes and beliefs about mathematics were administered at the beginning and end of the semester. Later studies on metacognition and affect were qualitative in nature. An attitude survey and measures of ability were used to select four to six students for in-depth observation and interviews. The interviews focused on students' use of a heuristic and their metacognitive control structures when solving problems.

Methods of Analysis

The assessment on students' concept of function was assessed independently by two researchers. The researchers then compared results and came to a consensus on dissenting items. Student justifications for an object qualifying or not qualifying as a function were categorized into 12 classes.

The quantitative research on affect used a nonequivalent control group design with initial and final surveys on attitude and belief. A multivariate analysis of variance with a covariate (MANCOVA) was used to test the hypotheses. The independent variable was treatment; the dependent variables were the seven subscale summed scores for the final attitude survey; and the covariates were the seven subscale summed scores for the initial attitude survey. Descriptive statistics were used to explore item-by-item comparisons between the control and experimental groups.

A qualitative case study design was used to explore the affective and cognitive questions addressed in later studies. The data consisted of interviews and attitude surveys. The interviews were analyzed using a coding scheme based on current theories of affect and function cognition. Two independent raters coded the interviews, using the coding scheme to perform a detailed analysis of student affect and cognition.

FINDINGS

Cognition: Function Concept

Function is a central concept in the study of algebra, particularly for an FPC that focuses on using functions as a model. But a student's concept image of function when beginning algebra was often a narrow view of function as a rule or a graph. The FPC moved students from this narrow relational view toward a view of function as a univalent relation. They also developed an understanding of function properties such as continuity, differentiability, and constant function. Qualitative data supported these quantitative findings, with students moving to a relational univalence concept based on a process interpretation of function. However, the properties of one-to-one and onto, as well as the concept of inverse, were still not well understood.

The majority of students entering college algebra held a strong process view of function when represented in either tabular or analytic (symbolic) form. However, they often viewed a graph as an object and a process. The FPC improved the student's process conception in all three representations. However, the object conception of function was much more problematic.

The ability to classify functions by their characteristics, such as linear versus nonlinear or increasing versus decreasing, suggests an object conception at the most rudimentary level. Finding a function's inverse, taking the composition of two functions, or translating a function suggests a more sophisticated object conception. Students' object conception of function in all but the graphic representation was weak at the beginning of the semester. The FPC improved students' object conception of function on both the rudimentary and sophisticated levels for all three representations. For example, those able to invert a function in tabular form increased from 3% to 46%, while in graphical form the figure rose from 9% to 51%. Those able to translate a function in analytic form increased from a low of 12.5% to 54%.

However, the percentage of those attaining at least a low-level object interpretation of function was still disappointing. Only about half of the students demonstrated any object conception of function. The qualitative data collected in later studies supported this finding. Keith, an African-American male, initially exhibited no conception of function beyond a relational view. He became so frustrated with the function questions presented without context that he stated, "I'm going to say no to all of those function things." However, when the function questions were contextualized in real-world situations, he exhibited a process conception of function in a tabular representation, interpolating to find an unknown population value; and he discussed the effect on the graph as an object. He was able to interpret the graph as both a process and an object, including describing the inverse of the graph via reflection in the line y = x. He recognized that there was a relationship represented by the graph, and since a function was a relation then there must be an analytic representation, but he could not determine the function. In summary, initially Keith held a relational view of function as a rule, operating from a process conception in tabular and analytic modes and from both a process and object conception in graphic mode.

After completing the FPC, Keith's function definition included all three representations, with a focus on function as a model for finding answers to real-world problems. Although he now recognized a graph and a table as representing a function, he still considered the analytic rule the best format for representing a function. Keith identified the constant relation y = 2 as a function, avoiding the misconception that a function must change dependent values. Keith used the univalence concept to successfully identify functions in both tabular and graphic form. He relied primarily on the vertical line test to determine univalance, but did discuss the concept of unique image via the table. He did not exhibit the continuity misconception or the differentiability misconception. He also used the concept of domain to argue that piecewise relations could be functions. This case study provides evidence that significant change did occur in the student's conceptual understanding. However, Keith's failure to interpret a formula or graph as a model of an application and his inability to complete a composition of functions task bring into question his level of understanding related to function as object.

Another measure of conceptual understanding is the ability to move between multiple representations of the concept. For functions this means translating among tabular, graphic, and analytic representations. Qualitative data indicated that initially students could translate from table to discrete graph using a process conception, but could not translate from either a table or graph to an analytic form.

After FPC exposure, students were proficient at converting from numeric to graphic, graphic to numeric, and analytic to numeric or graphic. However, they were not adept at converting from numeric to analytic or graphic to analytic. For example, Keith could use the process conception of function to translate from graphic to tabular, tabular to graphic, and analytic to both tabular and graphic. He could also use the process conception to interpret models of applications. But his lack of object conception greatly inhibited his ability to interpret models as objects.

Affective Issues

The initial quantitative studies into affect used a MANCOVA to analyze change in attitude and belief as measured by an initial and final attitude survey. We hypothesized that the reform curriculum would improve students' attitudes and beliefs concerning how algebra should be taught and learned. This was not supported. When interaction effects between the subscales were taken into account, there was no significant difference between students' attitudes in the FPC group and control group. The lack of significant difference may have been due to a variety of factors including:

1. limited exposure–one semester of experience with an FPC
2. the demanding nature of the FPC, which required conceptual understanding and the use of a CAS as a tool
3. adjusting to cooperative learning styles
4. teachers' attitudes and beliefs aligning with the curriculum

Observations of the ACT classrooms, as well as interviews with the ACT teachers, revealed an attitude on the part of the students that they were not "doing mathematics," since they were not always manipulating expressions. This may be one of the largest obstacles that any reform-oriented curriculum will have to overcome.

While the lack of statistical significance is disturbing, there is some evidence that the reform-oriented curriculum had positive effects on students' attitudes and beliefs. The treatment group had higher mean scores than the control group on 27 of the 40 survey items, and relatively equal mean scores (within 0.1) on seven more survey items. Thus the treatment group had lower mean scores than the control group on only 6 of 40 survey items. The treatment group had higher or equal mean scores than the control group on all items in the technology, applicability, and cooperative enterprise subscales. This indicated that students' attitudes toward using a CAS to discover mathematics improved with exposure to using Derive, students who focused on applications in mathematics appeared to find mathematics more useful, and students had more positive attitudes about learning mathematics via group projects.

The treatment students had higher problem-solving subscale means than the control students on four of six items and were equivalent on one item, indicating that their attitudes and beliefs about mathematics as problem-solving were more positive than their control counterparts. However, there were indications that students in the FPC group still preferred to avoid the emotion brought on by blockage when solving a problem for which they did not know a rule.

The conceptual understanding subscale means were greater for the treatment group on three of six items, equal on one, and less than the control group on two items; questions 12 (4.19 versus 4.35) and 15 (4.04 versus 4.39):

> Question 12: Even if I forget the formulas, I can still use the ideas to solve the problems.

> Question 15: My class focused too much on rules.

The FPC's focus on conceptual understanding versus rote use of rules did not appear to have the desired effect on students' belief about mathematics. Students in the reform curriculum still appeared to view mathematics as mostly rule-based.

The findings on the constructing knowledge and intrinsic motivation subscales indicated there was no difference between the experimental and control groups. One goal of the FPC was to move students toward a constructivist philosophy of learning mathematics. However, students in the treatment group were more reluctant than their counterparts in the control group to assume the responsibility for learning. It is essential that reform curricula require the student to assume this role, then provide the appropriate classroom environment to nurture its growth. Also, despite the long-term projects and open-ended exercises required by the FPC, treatment students still believed that mathematics concepts should be learned in a shorter amount of time than did the control group students. The FPC group appeared to be persisting in the belief that all mathematics problems should be solved in a short amount of time.

Later qualitative-based studies used in-depth interviews and observations to determine students' attitudes, beliefs, and emotions. Affective issues addressed in the interviews included:

1. Beliefs about mathematics as a subject of study and as an activity.
2. Beliefs about mathematical learning as related to ability and locus of control.

3. Attitudes toward mathematics and pedagogical elements of the FPC, such as applications motivating concepts, technology as a tool for constructing understanding, and social construction of knowledge.
4. Emotional reactions of frustration, fear, anger, etc.

For Kris, a caucasian female, mathematics was the manipulation of numbers and was related to low-level applications such as estimating your grocery bill. Her view of mathematics was predominantly static:

> Kris: There's rules for everything and there's ways to do it, and concepts. Math requires a good memory. I think that because there's so much to math.

> Interviewer: What is the most important thing for you to do so that you will remember the math?

> Kris: Probably a lot of repetition . . . Doing a lot of problems, and really grasping it.

Kris expressed a belief that mathematical ability was central to success, but effort did play a part. A poor self-image of her ability, combined with a static view of mathematics, resulted in an external locus of control. She had a negative attitude toward mathematics, an instantiation of poor self-esteem in her ability to do math, and negative experiences in previous mathematics courses. Her attitude toward math as solving real-world problems was negative, and she viewed conceptual understanding as remembering the correct algorithm. She had a positive attitude toward the possibility of using computers as a tool in exploring mathematics, but had no practical experience in doing so. She had a positive attitude toward working in small groups, because you learn different perspectives and ways to solve problems, but not in whole class participation. Kris held a negative attitude toward studying the history of mathematics to see its human face and writing in mathematics to improve conceptual understanding. The salient emotions exhibited by Kris were mathematics anxiety, excitement about using technology to explore mathematics, and excitement about cooperative work. She had a strong negative emotional reaction to reform-oriented curricula, based on her high school experience. She also discussed the emotion of frustration arising when she did not know how to complete a problem.

Changes in Kris's affective domain after participation in the FPC were significant. Kris's view of mathematics was more dynamic, since real-world applications and understanding concepts were now a focus for her. Self-image of her ability in mathematics was greatly improved: "I have done the best I could in this class . . . I understand math much better . . . I believe that I can do it." Her locus of control made a slight shift toward an internal locus, with effort taking on more importance. However, she still found mathematics to be difficult and not fun to do.

Kris's attitudes toward the pedagogical aspects of the FPC were maintained or improved. She now viewed real-world applications as much more than numbers applied in basic life skill situations. Her attitude toward technology as a tool for exploring mathematics was very positive: "Derive [the CAS used in the FPC], I think it's great; I mean the labs were." However, she had concerns about the automation of processes using a CAS leading to an erosion of basic skills: "I really don't know how to graph a function still because I always let Derive do it." She improved her attitude toward social construction of knowledge in a group setting, be it a large or small group. Her attitude toward history remained negative. But her attitude about writing to improve conceptual understanding improved, as long as the papers were limited to one or two pages.

With respect to emotions, Kris greatly reduced her mathematics anxiety due to her improved self-image concerning mathematical ability. She maintained her enthusiasm for using technology and cooperative learning. But perhaps the most telling change was in her emotions surrounding participation in a reform curriculum, which changed from strongly negative to positive. She endorsed the laboratories and stated, "I definitely feel better about myself when it comes to math."

Problem-Solving and Metacognition

Our current study focuses on the effect of the FPC on students' problem-solving and the metacognitive processes of self-regulation and self-monitoring. Four in-depth case studies are currently being analyzed. Each case study consists of in-class observations, written materials such as quizzes and exams, and three interviews with the following focuses:

Interview 1:   Compare the subject's natural and mathematical heuristics. What is a problem in the real world versus mathematics? What is the student's affect toward problem solving with respect to locus of control, reason for success, belief about mathematical problems, and emotion? Determine student's metacognition in problem solving.

Interview 2:   Midterm check on affect. How do students apply a heuristic for different problem types: solving polynomial equations, word problems with a known relationship, and modeling discrete data? Determine student's metacognition in problem solving by type.

Interview 3:   Final affect check.
How do students apply a heuristic in modeling a discrete data problem?
Focus is on student self-observation and discussion of heuristic and metacognitive aspects of problem solving.

While we are only at the beginning stages of analyzing the data, the following are some preliminary findings.

First, students hold a different conception of problem solving in mathematics versus problem solving in their everyday lives. Their heuristic or general method of attacking problems is often much more dynamic and self-actualized when working in non-mathematical realms. Second, formal mathematical heuristics are ignored even when explicitly taught. Students retain a simple heuristic: Read the question once, select a strategy (most likely the current topic being taught), and apply the strategy without monitoring progress. Third, in addition to poor self-monitoring, students seldom self-regulate; that is, they do not consider multiple strategies and try a new one when they are stuck. Rather, they either consult an authority or quit. Finally, students appear to apply different cognitive and metacognitive styles when solving different types of problems, such as finding the roots of an equation, solving an application using a known relationship or formula, and modeling a problem based on data analysis.

CONCLUSIONS

Issues of Affect

The FPC resulted in an impressive change in students' affect as related to mathematics. The students believed that they could do mathematics and that their success was due to their efforts. They saw mathematics as a useful tool for modeling real-world problems. Students moved from an external locus toward an internal locus, taking more responsibility for their learning. They also moved from a very static view of mathematics to a dynamic view, changing from a manipulation perspective to a conceptual understanding perspective. A CAS was seen as a useful tool in modeling and exploring mathematics. Students held positive attitudes about using applications to motivate the study of mathematics, moving from a low-level arithmetic view of the applicability of mathematics to a function modeling view.

Implications for Technology

We believe that student use of a CAS to explore and discover concepts and model real-world problems was central to the improvement in conceptual understanding we observed. The CAS provided the opportunity for students to construct some of the concepts for themselves by alleviating the manipulation and computation burden associated with generating examples and testing conjectures. The CAS also provided multiple representations of the function when students studied concepts and modeled problems, thus allowing them to attack the problems from numeric, graphic, and analytic approaches. With one representation supporting and enriching another, students gained a deeper understanding of the function concept. Finally, the computer allowed students to quickly and easily address "what if" type questions and conjectures generated by the student groups. This fostered an attitude of exploring, discovering, and confirming concepts, rather than passively receiving and rotely memorizing them.

Two concerns with using a CAS were cognitive overload and computer algebra usage. Derive was an easy-to-use CAS that greatly reduced student cognitive load with respect to using the software program. But even learning Derive while studying mathematics concepts overloaded some students. Teachers must be aware of the potential for cognitive overload when using technology and should provide explicit instruction on the use of the CAS. The introduction of new CAS commands and procedures should be done only when they are directly applicable to the mathematical topic being studied.

Using the computer algebra portion of a CAS was also a concern. We did not want to replace student competence in solving equations and manipulating expressions by calling up a CAS command they did not understand. Computer algebra commands were applicable when the focus was on the discovery of concepts using inductive reasoning. The CAS performed the manipulations so that the student could focus on the more abstract tasks of conjecturing and verification. A CAS was also applicable when modeling real-world data, allowing the student to focus on problem solving while the CAS implemented the function model. A CAS was particularly appropriate in the heuristic processes of exploring strategies, implementing the selected strategy, and extending the solution to related problems.

Implications for Instruction

Nemirovsky (1996) questioned the use of real-world problems in the functional perspective, dismissing three of four expectations of using this approach. The first expectation was that real-world problems enabled students to approach new mathematical concepts using ideas and situations that were familiar to them. While we do not discount Nemirovsky's concern that what is real is based upon the student's experience, we believe that real-world problems are closer to the student's experience than most well-formulated mathematical problems. Kris and Keith were very motivated by the modeling aspect of the FPC. This led to the second expectation, that formal definitions are natural and reasonable, being fulfilled to some extent. Both students came to believe that mathematics was more about concepts and less about rules. The function concept was seen as a natural and reasonable means for modeling real-world data, so students were motivated to strengthen their concept image of function.

Arguing that complexity is not an exogenous factor defined by the problem, we dismissed the third expectation that students learn to deal with complexity via real-world problems. Of this there is no doubt. Complexity is determined by the student interaction with the problem and with how the problem relates to his or her mental structures. We believe the notion that real-world complexity comes from working with messy data (meaning messy numbers) is a misnomer. The complexity of a real-world problem comes from dealing with extraneous information, selecting an appropriate modeling technique, and determining which model best matches the data. The process of modeling real-world data requires multiple representations of the function concept, as well as translating between those representations. Students in the FPC interacted with the problem because it was real. The complexity of modeling required them to engage their mental structures.

The fourth expectation is that real-world modeling helps students see the utility of mathematics. This is the expectation that underlies and motivates the other three. Kris had a negative attitude toward mathematics because it was difficult for her. She viewed mathematics as useful only in balancing her checkbook or calculating a grocery bill. After participating in the FPC, she still stated that mathematics was difficult for her; however, the utility of algebra in modeling real-world problems engaged her in learning. She saw a real-world usefulness in the function concept and put forth an effort to learn.

The FPC is about much more than a function approach with a focus on modeling real-world problems, and many questions about the FPC remain unanswered. How can we motivate students to construct a stronger object conception of function? What are the effects on equation concepts of introducing functions before equations? Is cognitive overload occurring due to the concept of modeling and adding a CAS to the curriculum? How do we get students to reconcile discrete and continuous representations of models? What characteristics must an instructor using an FPC exhibit? While these questions require further investigation, our research supports the use of an FPC employing a CAS as a tool in the instruction of algebra.

REFERENCES

Bednarz, N., Kieran, C., & Lee, L. (1996). *Approaches to algebra: Perspectives for research and teaching.* Dordrecht: Kluwer Academic Publishers.

Bell, A. (1996). Problem-solving approaches to algebra: Two perspectives. In N. Bednarz, C. Kieran, & L. Lee (Eds.), *Approaches to algebra: Perspectives for research and teaching* (pp. 167-186. Dordrecht: Kluwer Academic Publishers.

Blum-Anderson, J. (1992). Increasing enrollment in higher-level mathematics classes through the affective domain. *School Science & Mathematics, 8*, 433-436.

Fey, J., & Heid, K. (1995). *Concepts in algebra: A technological approach.* Dedham, MA: Jason Publications.

Hart, L.E. (1989). Describing the affective domain: Saying what we mean. In D.B. McLeod & V.M. Adams (Eds.), *Affect and mathematical problem-solving: A new perspective* (p. 37). New York: Springer-Verlag.

Heid, K. (1996). A technology-intensive functional approach to the emergence of algebraic thinking. In N. Bednarz, C. Kieran, and L. Lee (Eds.), *Approaches to algebra: Perspectives for research and teaching* (pp. 239-256). Dordrecht: Kluwer Academic Publishers.

Heid, K (Ed.). (1995). *Algebra in a technological world.* Reston, VA: National Council of Teachers of Mathematics. Dordrecht.: Kluwer Academic Publishers.

Mandler, G. (1989). Affect and learning: Causes and consequences of emotional interactions. In D.B. McLeod & V.M. Adams (Eds.), *Affect and mathematical problem-solving: A new perspective* (p 3). New York: Springer-Verlag.

Mason, J. (1996). When is a problem? Questions from history and classroom practice in algebra. In N. Bednarz, C. Kieran, & L. Lee (Eds.), *Approaches to algebra: Perspectives for research and teaching* (pp. 187-193). Dordrecht: Kluwer Academic Publishers.

Mayes, R. (1998). ACT in algebra: Student attitude and belief. *The International Journal of Computer Algebra in Mathematics Education, 5*, 17-38.

Mayes, R., & Lesser, L. (1998). *Act in algebra.* Boston: WCB McGraw-Hill.

McLeod, D.B. (1988). Affective issues in mathematical problem-solving: Some theoretical considerations. *Journal for Research in Mathematics Education, 25*(6), 637-647.

McLeod, D.B. (1992). Research on affect in mathematics education: A reconceptualization. In D.A. Grouws (Ed.), *Handbook of research on mathematics teaching and learning* (p. 575). New York: Macmillan.

McLeod, D.B. (1994). Research on affect and mathematical learning in the JRME: 1970 to the present. *Journal for Research in Mathematics Education, 25*(6), 637-647.

Middleton, J.A. (1995). A study of intrinsic motivation in the mathematics classroom: A personal constructs approach. *Journal for Research in Mathematics Education, 26*(3), 254.

Moschkovich, J., Schoenfeld, A., & Arcavi, A. (1993). Aspects of understanding: On multiple perspectives and representations of linear relations and connections among them. In T.A. Romberg, E. Fennema, & T. P. Carpenter (Eds.), *Integrating research on the graphical representation of functions* (pp. 69-100). Hillsdale, NJ: Lawrence Erlbaum.

Nemirovsky, R. (1996). A functional approach to algebra: Two issues that emerge. In N. Bednarz, C. Kieran, & L. Lee (Eds.), *Approaches to algebra: Perspectives for research and teaching* (pp. 295-313). Dordrecht: Kluwer Academic Publishers.

Novak, J., & Gowin, B. (1984). *Learning how to learn.* New York, NY: Cambridge University Press.

O'Callaghan, L. (1998). Computer-intensive algebra and students' conceptual knowledge of functions. *Journal for Research in Mathematics Education, 29*(1), 21.

Radford, L. (1996). Some reflections on teaching algebra through generalization. In N. Bednarz, C. Kieran, & L. Lee (Eds.), *Approaches to algebra: Perspectives for research and teaching* (pp. 107-111). Dordrecht: Kluwer Academic Publishers.

Sfard, A. (1995). The development of algebra: Confronting historical and psychological perspectives. *Journal of Mathematical Behavior, 14*, 15-39.

Slavit, D. (1994). *The effect of graphing calculators on students' conceptions of function.* Paper presented at the Annual Meeting of the American Educational Research Association, New Orleans, LA.

Stage, F.K., & Kloosterman, P. (1992). Measuring beliefs about mathematical problem-solving. *School Science & Mathematics, 3*, 109.

Vinner, S., & Dreyfus, T. (1989). Images and definitions for the concept of function. *Journal for Research in Mathematics Education, 20*(4), 356-366.

Wheeler, D. (1996). Backwards and forwards: Reflections on different approaches to algebra. In N. Bednarz, C. Kieran, & L. Lee (Eds.), *Approaches to algebra: Perspectives for research and teaching* (pp. 317-325). Dordrecht: Kluwer Academic Publishers.

Larry J. Stephens
John Konvalina

Factors Influencing Success
in Intermediate Algebra

SUMMARY. The purpose of this study was to investigate the following factors with respect to their influence on success in an intermediate college algebra course: short weekly quizzes, computer algebra software projects, and a practice comprehensive final exam. The results suggest that all three factors significantly influence success in an intermediate algebra course. Success was measured by achievement on an independent final exam. *[Article copies available for a fee from The Haworth Document Delivery Service: 1-800-342-9678. E-mail address: <getinfo@haworthpressinc.com> Website: <http://www.HaworthPress.com> © 2001 by The Haworth Press, Inc. All rights reserved.]*

KEYWORDS. Intermediate algebra, computer algebra software, warm-up final exam, weekly tests, booster points, step-wise regression, correlation

Whether at the high school or college level, the teaching of algebra is considered to be a very difficult task. It is often difficult to motivate students to practice the techniques used in factoring, simplifying expressions, and all

LARRY J. STEPHENS is Professor, Department of Mathematics, University of Nebraska, Omaha, NE 68182 (E-mail: Stephens@Unomaha.edu).
JOHN KONVALINA is Professor, Department of Mathematics, University of Nebraska, Omaha, NE 68182 (E-mail: Johnkon@Unomaha.edu).

[Haworth co-indexing entry note]: "Factors Influencing Success in Intermediate Algebra." Stephens, Larry J., and John Konvalina. Co-published simultaneously in *Computers in the Schools* (The Haworth Press, Inc.) Vol. 17, No. 1/2, 2001, pp. 77-84; and: *Using Information Technology in Mathematics Education* (ed: D. James Tooke and Norma Henderson) The Haworth Press, Inc., 2001, pp. 77-84. Single or multiple copies of this article are available for a fee from The Haworth Document Delivery Service [1-800-342-9678, 9:00 a.m. - 5:00 p.m. (EST). E-mail address: getinfo@haworthpressinc.com].

77

the various concepts encountered in algebra. Without the repetition of techniques provided by working several different problems, it is not easy to master the concepts of algebra. Stephens and Konvalina (1999) found that the introduction of computer algebra software into intermediate and college algebra courses motivated students to work more problems. The students were assigned a variety of algebra problems to be solved using computer algebra software. They were encouraged to check, by hand, the solutions to assigned problems obtained by using the computer software package Maple (Harper, Wooff, & Hodgkinson, 1991; Heck, 1993). The results of an opinionnaire administered at the end of the courses showed the students were positive about the use of Maple in the courses.

Previous studies have used computer algebra systems to enhance the learning of mathematics. Computer algebra systems such as Derive, Maple, and Mathematica (Harper, Wooff, & Hodgkinson, 1991) are powerful software packages that are available for use in various mathematics courses. Derive was used as a demonstration device in the classroom as well as being available for student use in a computer laboratory (Mayes, 1995). Students in the experimental group outperformed those in the traditional lecture classes with respect to inductive reasoning, visualization, and problem solving.

Yerushalmy and Gilead (1997) discuss the use of the software package The Function Supposer: Functions and Graphs with eighth-grade students when studying the solutions of equations. The students used the software as a tool in their efforts at solving linear as well as nonlinear equations.

The software packages Derive and Gcal (a graphic introduction to calculus) were used in a study involving engineering mathematics (Lawson, 1995). The group using the software showed a positive benefit when compared to another group who received a traditional lecture. Cheung (1996) discusses the use of Maple (Heck, 1993) in a number theory course taken by prospective teachers in primary and secondary schools.

It is surprising that very little research has been done on studying the effects of computer algebra systems, such as Maple, in algebra courses at the high school and college levels. During the past decade, the impact of computer technology on education, especially mathematics education, has been enormous. Educators must integrate the technology into their courses as well as determine the effectiveness on the learning process.

The purpose of this investigation was to compare the effects of the following factors upon achievement in an intermediate, college-level algebra course: short weekly quizzes, extra credit Maple assignments related to various topics in the course, and a take-home comprehensive pre-final examination. The purpose of the short weekly quizzes was to motivate the student to keep up in the course (not fall behind). The purpose of the Maple assignments was to introduce computer technology and, hopefully, increase student interest in learning algebra. The purpose of the comprehensive pre-final examination was to assist students in organizing their preparation for the in-class comprehensive final examination. The effect of computer technology on achievement in intermediate algebra was of particular interest.

METHOD

Participating in the study were 30 students enrolled in an intermediate, college-level algebra course. Thirteen weekly tests worth 10 points each were given, as well as a midterm and a final worth 100 points each. Associated with each weekly test was a weekly test booster. The weekly booster consisted of five relevant problems to be worked using the computer algebra package Maple. The booster worked in the following way. Each problem solved correctly using Maple would add a point to the weekly test score. However, the weekly test score could not exceed 10. So, for example, if a student scored 7 on the weekly test and got all five booster problems correct, the weekly test score would be adjusted to 10, but not to 12. The test booster was voluntary. Table 1 shows some of the algebra topics covered by the booster problems, the Maple commands, and the answers given by Maple.

A take-home, comprehensive pre-final worth 25 points and consisting of 25 questions covering the complete course, called the final warmup, was

TABLE 1. Some Maple Commands Used in the Study

Algebra Topic	Maple Command	Maple Answer
Solving linear equations	> solve (P = 2*L + 2* W,W);	$1/2 P - L$
Solving linear inequalities	> solve (-3^* (x + 4) + 2 > = 8 $-$x);	($-$infinity, -9)
Slope of a line	> with (geometry): > line (L,3* y + 2* x = 9,[x,y]); > slope(L);	$-2/3$
Systems of equations	> solve ({2*x + 3*y = $-$6,x $-$ 3*y = 6, {x,y});	{y = -2, x = 0}
Simplify expression	> simplify ((2^3*x^ ($-$ 2))^($-$2));	$1/64 \, x^4$
Multiply complex numbers	> (3 + 5*I)*(4 $-$ 2*I);	22 + 14I
Solve polynomial equation	> solve(x^4 $-$ 13*x^2 + 36);	2, 3, -3, -2
Multiply polynomials	> expand((3*m $-$ 5)*(3*m^3 $-$ 2*m^2 + 4));	$9m^4 - 21m^3 + 10m^2 + 12m - 20$

administered a week prior to the final and was due the last day of class. Finally, a departmental final consisting of 25 questions and counting 100 points was given in the course.

DESIGN AND STATISTICAL ANALYSIS

The raw weekly test scores were summed for each student and converted to a percentage. Similarly, the percentage of points obtained on the Maple boosters, the final warmup, the midterm, and the final were obtained for each student. Descriptive statistics were computed for each of these variables. In order to access the relationship among the variables, correlation coefficients were computed, along with their p-values. Finally, a step-wise regression using the final as the response and the other variables as independent variables was performed.

RESULTS

Table 2 gives the mean, minimum, and maximum for the five variables in the study. The descriptive statistics indicate that these students were poor achievers in algebra. They scored an average of 60% on their weekly tests, 70% on the midterm, and only 56% on the comprehensive final examination.

Furthermore, the scores on the booster and the warmup indicate that, even if given a chance to improve their test performances, many students were not willing to spend the time doing the Maple assignments or to put much time into the practice final.

Table 3 gives the correlation coefficients for the 10 pairs of variables. Of interest is that achievement on the final examination is significantly correlated with performance on the weekly tests, the number of Maple problems correctly solved, and the performance on the practice final examination.

Table 4 gives the results of a step-wise regression where the response is the score on the final examination and the independent variables are weekly

TABLE 2. Descriptive Statistics for Variables

Variable	Mean	Minimum	Maximum
Final	56	20	96
Midterm	70	40	96
Booster	44	0	78
Weekly	60	34	78
Warmup	75	0	100

TABLE 3. Correlation Coefficients and *p*-Values for Variables

	Final	Midterm	Booster	Weekly
Midterm	0.574 0.001			
Booster	0.442 0.014	0.371 0.044		
Weekly	0.605 0.000	0.555 0.001	0.228 0.225	
Warmup	0.394 0.031	0.444 0.014	0.376 0.041	0.481 0.007

TABLE 4. Results of Step-Wise Regression

Step-Wise Regression

F-to-Enter: 4.00   *F*-to-Remove: 4.00
Response is final on 3 predictors, with *N* = 30

Step	1	2
Constant	10.981	8.032
weekly	0.74	0.65
T-Value	4.02	3.68
booster		0.189
T-Value		2.22
S	12.2	11.4
R-Sq	36.64	46.41

test scores, booster points, and score on the practice final. The *F* value to enter and remove is 4.

The first variable to step in was the performance on the weekly tests. This variable accounted for about 37% of the variation in the scores on the final. The best two variables were weekly test performance and booster test points, accounting for over 46% of the variation in scores on the final. It is also of interest to look at the simple regression plots of final examination score versus the independent variables one at a time. These plots are shown in Figures 1, 2, and 3.

DISCUSSION

The results of this study seem to reinforce that old saying, "You can lead a horse to the water, but you can't make him drink." For those students willing

FIGURE 1. Final exam scores versus number of booster points.

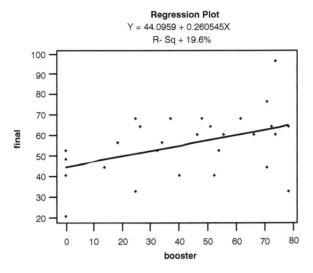

Regression Plot
Y = 44.0959 + 0.260545X
R- Sq + 19.6%

FIGURE 2. Final exam scores versus weekly test scores.

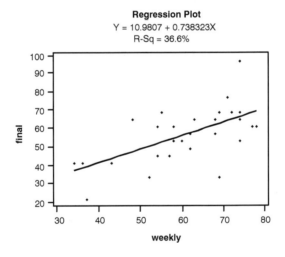

Regression Plot
Y = 10.9807 + 0.738323X
R-Sq = 36.6%

to keep up weekly and whose weekly test scores were high, their achievement in algebra was generally better than for those who did not. The more Maple-assigned problems worked by the student, the better the achievement. Four of the 30 students did not work a single one of the 65 booster problems. Their scores on the final examination were 20, 40, 48, and 52. There was also a

FIGURE 3. Final exam scores versus final warmup test scores.

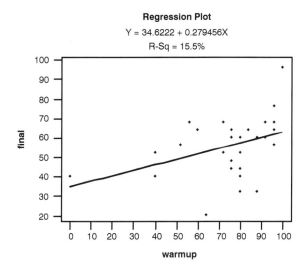

significant positive correlation between the performance on the practice final and the final examination score in the course. It appears that weekly testing, integration of computer algebra software into the course, and a practice final examination are worthwhile efforts. Clearly, more research is needed to determine the long-term impact of the computer technology in intermediate algebra.

ACKNOWLEDGMENT

The data analysis in the tables and the graphs in Figures 1, 2, and 3 were produced by the use of the software package MINITAB™.

REFERENCES

Cheung, Y. L. (1996). Learning number theory with a computer algebra system. *International Journal of Mathematical Education in Science and Technology,* 27(3), 379-385.

Harper, D., Wooff, C., & Hodgkinson, D. (1991). *A guide to computer algebra systems.* New York: Wiley.

Heck, A. (1993). *Introduction to Maple.* New York: Springer-Verlag.

Lawson, D. A. (1995). The effectiveness of a computer-assisted learning programme in engineering mathematics. *International Journal of Mathematical Education in Science and Technology, 26*(4), 567-574.

Mayes, R. L. (1995). The application of a computer algebra system as a tool in college algebra. *School Science and Mathematics, 95*(2), 61-67.

Stephens, L. J., & Konvalina, J. (1999). The use of computer algebra software in teaching intermediate and college algebra. *International Journal of Mathematical Education in Science and Technology, 30*(4), 483-488.

Yerushalmy, M., & Gilead, S. (1997). Solving equations in a technological environment. *The Mathematics Teacher, 90*(2), 156-162.

Leping Liu
Rhoda Cummings

A Learning Model That Stimulates Geometric Thinking Through Use of PCLogo and Geometer's Sketchpad

SUMMARY. PCLogo and Geometer's Sketchpad are powerful tools that may be used by mathematics teachers who want to integrate technology with geometry instruction in the elementary classroom. The purpose of this study was to examine the usefulness of PCLogo and Geometer's Sketchpad to stimulate thinking about geometric concepts in elementary age children. We used a collective case study design that included four cases, two girls (ages 8 and 10) and two boys (ages 10 and 11). All participants were trained to use PCLogo and Geometer's Sketchpad to construct geometric shapes and measure the attributes of the shapes. After the training, participants used these technologies as tools to stimulate thinking about geometric concepts. As a result of our observations of children's performances as they reasoned about geometric concepts, we developed a learning model for teaching children about geometry. *[Article copies available for a fee from The Haworth Document Delivery Service: 1-800-342-9678. E-mail address: <getinfo@haworthpressinc. com> Website: <http://www.HaworthPress.com> © 2001 by The Haworth Press, Inc. All rights reserved.]*

LEPING LIU is Assistant Professor, Department of Reading, Special Education and Instructional Technology, College of Education, Towson University, Towson, MD 21252 (E-mail: lliu@towson.edu).
RHODA CUMMINGS is Professor, Department of Counseling and Educational Psychology, University of Nevada, Reno, NV 89557-0029 (E-mail: cummings@unr.edu).

[Haworth co-indexing entry note]: "A Learning Model That Stimulates Geometric Thinking Through Use of PCLogo and Geometer's Sketchpad." Liu, Leping, and Rhoda Cummings. Co-published simultaneously in *Computers in the Schools* (The Haworth Press, Inc.) Vol. 17, No. 1/2, 2001, pp. 85-104; and: *Using Information Technology in Mathematics Education* (ed: D. James Tooke and Norma Henderson) The Haworth Press, Inc., 2001, pp. 85-104. Single or multiple copies of this article are available for a fee from The Haworth Document Delivery Service [1-800-342-9678, 9:00 a.m. - 5:00 p.m. (EST). E-mail address: getinfo@haworthpressinc.com].

85

KEYWORDS. PCLogo, Geometer's Sketchpad, geometry, learning model, mathematical thnking skills, math software, abstract reasoning, problem solving, geometric principles, concrete-abstract process, abstract-concrete process

Mathematics teachers constantly face the challenge of how to use technology as an instructional tool. Teachers of mathematics also are expected to adhere to the *Curriculum and Evaluation Standards for School Mathematics* (NCTM, 1989) and the *Professional Standards for Teaching Mathematics* (NCTM, 1991). Not only do these standards emphasize the development of students' mathematical thinking abilities, they also imply that instruction in geometry should be included in the elementary school curriculum (Geer, 1999). The standards specific to instruction in geometric thinking are included in the *Principles and Standards for School Mathematics 2000* (NCTM, 1999). In response to these standards, new and innovative computer software has been designed to enhance mathematical thinking in children (Davis & Hersh, 1981; Abramovich & Nabors, 1997; Vacc & Bright, 1999). In view of the demands placed on teachers by the new math standards to teach geometry at the elementary level, an important goal for mathematics teachers should be to use new software technology to advance children's thinking about geometry.

A number of studies have investigated the use of instructional technology to improve students' mathematical thinking skills. Findings of these studies suggest that use of well-designed mathematical software programs that incorporate effective procedures for teaching problem solving can help children develop the abstract reasoning skills required for high levels of mathematical thinking, such as thinking about abstract geometric concepts.

BACKGROUND

van Hiele's Three Levels of Geometric Thinking

Traditional theories of cognitive development suggest that learning may not occur unless instruction is appropriate to children's level of cognitive development (Piaget, 1936; van Hiele, 1986). Thus, in geometry learning, which requires abstract thinking abilities, children theoretically should not be able to understand certain geometric principles until their cognitive development is at the formal, abstract level. According to van Hiele's geometric thinking hierarchy, children's learning and thinking about geometry changes as they advance through three levels (van Hiele, 1986, 1997; Fuys, Geddes, & Tischler, 1988):

1. Visual level
2. Descriptive/analytic level
3. Abstract/relational level

At Level 1, the visual level, children identify and think about geometric shapes based on the visual appearance of shapes and according to their similarities to real-world objects (van Hiele, 1999). For example, children will recognize a shape as a circle because it looks like a round plate. At level 2, the descriptive/analytic level, children think about geometric shapes according to their concrete characteristics (Liu & Cummings, 1997; van Hiele, 1999). At this stage, for instance, a shape is a rectangle, not because it looks like a door, but because it has two long sides, two short sides, and four right angles. At Level 3, the abstract/relational level, children are able to integrate the visual information gained in Level 1 and their understanding of the characteristics of shapes gained in Level 2 to construct an abstract concept of geometric principles (Liu & Cummings, 1997; van Hiele, 1999). Thus, children at this level of thinking understand and can explain why the sum of the angles of any quadrilateral must be 360 degrees.

Although the van Hiele model describes the *progression* of children's thinking from one level to the next, the model does not describe the *processes* that are involved in advancing thinking through the three levels (Liu, 1999). It is important, therefore, to understand how a child's thinking progresses from one level to the next and to identify the thinking processes that must be in place before the transition can occur.

Two Processes of Geometric Thinking

The processes involved in advancing geometric thinking require both *concrete thinking* and *abstract thinking*. In much of the cognitive development literature, concrete thinking and abstract thinking are described as two separate stages of thinking (Swan, 1993), similar to van Hiele's three discrete levels. For example, in his classical theory of cognitive development, Piaget (1971) makes a distinction between concrete thinking and formal thinking. Concrete-thinking children can think logically and solve mathematical problems. However, concrete thought can only solve problems that exist in the present and that can be represented through manipulation of physical objects. For example, the concrete-thinking child cannot solve an abstract verbal problem, such as "John is taller than Susan, and Susan is taller than Bill. Who is the tallest of the three?" However, when concrete materials are used, such as blocks of different sizes and colors, the child can easily solve a similar problem such as: "The red block is bigger than the yellow block, and the yellow block is bigger than the blue block. Which is the biggest block?" In comparison, formal abstract thinking stems from mental rather than physical manipulations, and problem-solving skills do not depend on concrete experiences.

Liu and Cummings (1997) have described two thinking *processes* that are essential for geometry learning and that advance movement through van Hiele's hierarchy:

1. Concrete-abstract process (CA)
2. Abstract-concrete process (AC)

The CA process accounts for advancement through the three levels of the van Hiele hierarchy. The CA process begins with children's initial sensation of concrete objects and experiences in the physical world (Liu & Cummings, 1997). Once these physical stimuli have been detected by the sensory system, their particular qualities and characteristics are identified and interpreted through perception. This process ends as the individual formulates concepts, ideas, or laws about what was sensed and perceived, extracting an abstract concept of the concrete experience. The CA process also can be conceived of as a process of inductive thinking–reasoning from particular facts (geometric shapes, measurements, etc.) to a general conclusion about concepts, ideas, or laws (geometric concepts or rules). In geometric learning, the CA process leads thinking through the van Hiele hierarchy and stimulates transitions between each of the three levels.

Once children go through the concrete-abstract thinking process, their thinking about geometry progresses to the third level of van Hiele's hierarchy, the abstract/relational level. According to Liu and Cummings (1997):

> Once children have developed these concepts, they have developed a geometry schema that contains all of the rules and relationships that have been learned during the process . . . Moreover, once children develop these concepts, they can think about geometry in more complex ways than if they simply had memorized rules about the characteristic qualities of geometric shapes and figures. (p. 101)

In contrast to van Hiele's scheme, the abstract/relational level is not the highest level at which children can think about geometry. Once they reach this level, children are ready to move to an even higher level of thinking, that of *abstract-concrete* (AC) thinking, which allows them to *apply* their newly learned concepts (Liu & Cummings, 1997). The process of abstract-concrete (AC) thinking is not simply the reverse of concrete-abstract thinking. Instead, it is a higher form of thinking that depends on more advanced abstract and logical reasoning abilities but is grounded in the concepts and rules derived from concrete-abstract learning (Liu & Cummings, 1997). In other words, it is not enough just to think abstractly about geometric principles; the individual must be able to apply this thinking to solve actual problems.

The AC process also can be conceived of as a process of deductive thinking–reasoning from general to specific. In geometry learning, the application of deductive reasoning in the problem-solving process may occur through a sequence of steps. One model that describes this process is the six-step problem-solving model proposed by Hayes (1989). The six steps include:

1. Identifying the problem
2. Representing the problem
3. Planning the problem
4. Executing the plan
5. Evaluating the plan
6. Evaluating the solution

The Hayes sequence depicts the way information is handled during the problem-solving process and requires children to divide a single geometric problem into a number of smaller, more basic problems. Children use steps one through four to solve each smaller problem. Then, after evaluating the answers to each smaller problem, they configure their solutions into a final solution to the larger problem in steps 5 and 6.

In summary, concrete-abstract (CA) processes are involved as children's geometric thinking advances from the visual level to the abstract/relational level in van Hiele's hierarchy. Once they reach the abstract thinking level, children apply abstract-concrete (AC) thinking processes to solve concrete geometric problems. Application of CA and AC processes results in inductive and deductive reasoning skills, respectively.

Technology Tools

As previously noted, mathematics teachers are expected to integrate new technologies into classroom teaching to improve students' mathematical and geometric thinking skills. Two technology tools that are useful for accomplishing this purpose are PCLogo, a computer programming language, and Geometer's Sketchpad, a software package designed to enhance children's understanding of principles of geometry.

Like many computer languages, Logo provides users with the capability to create attractive graphic images, perform calculations, maintain and update data, and even create sounds and play music. In a Logo environment, even children can perform these tasks and have fun at the same time. However, there is one important difference between Logo and other languages. Logo was uniquely developed by Seymour Papert (an MIT scientist who studied with Piaget) as a conceptual framework for understanding children's construction of knowledge about mathematics and problem solving (Clements & Meredith, 1993; Liu & Cummings, 1997; Papert, 1980).

A number of studies have investigated the effectiveness of Logo as a concrete context for facilitating children's abstract reasoning about geometric problems (Geddes, 1992; Weaver, 1991). Findings of these studies suggest that Logo can be effective in motivating children to learn through exploration and discovery. Through these processes, children come to understand that there may be several solutions to any problem. As children experiment and look for different solutions to a task, they use different thinking skills and construct knowledge in different ways. By observing children as they experiment with Logo commands and develop procedures for solving geometric problems, it is possible to gain insight into their thinking and learning processes.

Geometer's Sketchpad is a useful tool for advancing children's thinking through van Hiele's hierarchy because it allows them to investigate geometric concepts and discover relationships among these concepts (Key Curriculum, 1999; Pokay & Tayeh, 1997). The Sketchpad provides users with the capability to draw, measure, calculate, and script geometric shapes and figures. Thus, it is a powerful tool for visually presenting geometric concepts to children and for allowing them to construct points, lines, and circles using constraints. For example, a user can constrain a point to be the midpoint of a line segment, set one line to be parallel to another, fix a circle's radius equal to a given length, and construct a graph of geometric relationships. Moreover, if any part of a geometric shape is transformed, all related parts change accordingly, allowing immediate observation of geometric relationships. These visual effects provide the concrete information that, when children sense it repeatedly, can be generalized into abstract geometric principles.

PURPOSE OF THE STUDY

The purpose of this case study was (a) to examine the effectiveness of PCLogo and Geometer's Sketchpad in stimulating geometric thinking in children, and (b) to identify the thinking processes used by children as they advance through van Hiele's geometric thinking hierarchy.

METHOD

This study was an *instrumental case study*, which is defined as:

> examining a particular case to provide insight into an issue or refinement of theory. The case plays a supportive role, facilitating our understanding of something else. When an instrumental study is extended to several cases, it is called a *collective case study*. (Denzin & Lincoln, 1994, p. 237)

We observed four children engaged in the process of learning about geometric concepts. We attempted (a) to determine how each child used PCLogo and Geometer's Sketchpad to learn geometric concepts, and (b) to identify the cognitive processes that were used to learn these concepts. As a result of our observations, we developed a learning model to teach abstract geometric principles to elementary age children.

Subjects

Four children participated in the study, two girls (ages 8 and 10) and two boys (ages 10 and 11). None of the children had previously used PCLogo or Geometer's Sketchpad. During an initial training session, we gave each child specific, step-by-step instructions for using Logo and Geometer's Sketchpad. After completion of the training session, the children practiced what they had learned by using Logo to create shapes and Geometer's Sketchpad to construct geometric figures and measure the various line lengths and angles of these figures. After completion of the training and practice sessions, the children had gained sufficient knowledge and skills to use these two tools to learn abstract geometric concepts.

Procedures

The learning model. In this study, not only were we interested in *what* the participants learned, but we also wanted to know *how* they learned by carefully observing them as they thought about and performed each step in the learning process. If we could identify the thinking processes used by children to learn about abstract geometric principles, we could develop a learning model to teach these principles to elementary age children. To structure our observations, we asked the following questions:

1. How does a student learn geometric concepts?
2. How does a student apply thinking skills to learn geometric concepts?
3. How does a student use PCLogo and Geometer's Sketchpad to learn geometric concepts?

Underlying assumptions of the learning processes. According to our theory of concrete-abstract (CA) and abstract-concrete (AC) learning processes (Liu & Cummings, 1997), we made two assumptions. First, we assumed that the participants had performed concrete-abstract thinking and used the CA process to learn geometric concepts if they could:

1. use PCLogo commands to construct a geometric shape, analyze the characteristics of the basic components of the shape, and summarize the definition of the shape correctly; and
2. use Geometer's Sketchpad to measure the basic components of a shape (lines and angles), analyze the characteristics of these components, and correctly summarize these characteristics into a geometric concept.

A geometric concept incorporates within it several basic components. For example, the basic components of the concept "rectangle" are the sides, the length of the sides, the angles, the degree of the angles, and the numbers of the sides and angles. These components constitute the "concrete" parts of a rectangle, which can be analyzed by (a) observing their similarities, (b) noticing their differences, and (c) discovering certain rules that apply when changes are made to properties of any of the components (such as lengthening a side or changing the degree of an angle). Eventually, an abstract concept of "rectangle" can be summarized from analysis of the characteristics of the concrete components of any geometric shape. These are the processes involved in development of concrete-abstract (CA) thinking.

Second, we assumed that children had performed abstract-concrete thinking and used the AC process to apply geometric concepts, if they could:

1. analyze the characteristics of a new shape, write the PCLogo code to produce the new shape, and correctly summarize the definition of the shape; and
2. analyze the characteristics of a new shape, conduct the procedures to measure the components of the shape with Geometer's Sketchpad, and correctly summarize the shape's conceptual characteristics.

These are the processes involved in abstract-concrete (AC) thinking, which is the *problem-solving process* involved in concrete application of an abstract concept. Children begin the process by analyzing the characteristics of a geometric shape, using the abstract knowledge they obtained in the CA process. Notice that there is a difference between analysis of characteristics in the CA process and in the AC process. In the CA process, children produce a geometric shape *in response to* a set of instructions that have been provided for them. In the AC process, children *formulate* the instructions to produce a geometric shape based on their understanding of the characteristics of that shape.

Learning tasks to determine if assumptions of the learning processes are met. To examine participants' CA and AC thinking processes, we designed four geometric tasks:

Task One (Logo with CA process): Given the Logo code "repeat 4 [forward 100 right 90]," ask the child to:

1. use PCLogo to produce the shape (explain that this shape is a "square");
2. use the Logo code to replace side length 100 with three different numbers, and produce the three shapes; and
3. summarize the definition of "square."

Task Two (Logo with AC process): Given a "rectangle" shape with 100 as the short side and 150 as the long side, ask the child to:

1. compare this shape with a square in terms of its sides and angles;
2. write the Logo code to produce the rectangle, and then produce two other rectangle shapes with different lengths and widths;
3. summarize the definition of "rectangle."

Task Three (Geometer's Sketchpad and CA process): Given a triangle, ask the child to:

1. measure each angle and calculate the sum degree of the three angles;
2. drag one angle to change the length of the sides, or the degree of the angles of the triangle, and write down the change of each angle and the sum degree of the three angles; and
3. draw a conclusion about the sum degree of three interior angles of a triangle.

Task Four (Geometer's Sketchpad and AC process): Repeat the method used in Task Three, to summarize the sum degree of angles of any quadrilateral shapes.

Tasks One and Three require concrete-abstract thinking, and Tasks Two and Four require abstract-concrete thinking. The following criteria were used to evaluate the participants' performance on each of these tasks:

1. The definitions or conclusions should be correctly described.
2. The Logo code should be correct and specific to the tasks.
3. The measurements and calculations should be accurate.
4. The procedures should be in a logical sequence.

Observations

We observed each child's performance and took detailed notes on the procedures used to complete the four tasks. The procedures included the Logo code used, the alternative angles or sides produced, and the measurements and calculations performed. Using the CA and AC thinking/learning processes as the framework for our observations, we focused on whether children's performances on each task were consistent with or different from the assumed processes. We also noted whether a child learned independently without directions or needed specific directions for completing the task.

THE CASES AND FINDINGS

The summary of our observations is structured according to (a) the type of training required to teach each child how to use PCLogo and Geometer's Sketchpad, (b) each child's performance on each of the four tasks, and (c) findings. In all four cases, the children learned by following the same procedures:

1. learning PCLogo,
2. working on Tasks One and Two,
3. learning Geometer's Sketchpad, and
4. working on Tasks Three and Four.

Case One

The participant was an eight-year-old girl, a third-grade student in a public school. Her teachers and parents reported that her classroom achievement ranged from average to above average. Her only previous experience with computers was to play games. However, she was curious about and interested in learning PCLogo and Geometer's Sketchpad. Although we had prepared step-by-step written instructions for using these programs, the participant could not read or understand the materials. We then demonstrated the software to her and showed her how to work through each step in the program. She followed the steps correctly and learned the necessary skills for solving all four tasks. We concluded from our observations of her performance during the training process that she learned well with visual examples and concrete verbal directions. Therefore, we explained the requirements and demonstrated the processes for solving each of the four tasks to her before she began working on them.

In Task One, she successfully produced the square shape with the given code as in Figure 1-A (1-B and 1-C refer to Cases Two and Three).

Next, we instructed her to change side length 100 to side lengths 150, 200, and 300, respectively. For each change in side length, she wrote the code correctly and produced the three squares. However, she had difficulty performing step 3, summarizing the definition of "square." To help her with this, we constructed a worksheet in which she could write down the numbers referring to the sides and angles of each of the four shapes (see Table 1). When she looked at the side lengths and angle degrees for each shape, she was able to summarize correctly that all four shapes had "four equal-length sides and four 90-degree angles." Thus, with guided assistance, the child eventually demonstrated performance of the concrete-abstract (CA) thinking/learning process.

Because of her learning experiences in Task One, the girl experienced greater success in completion of the steps in Task Two. She first compared

FIGURE 1. Logo and CA process (code to shape).

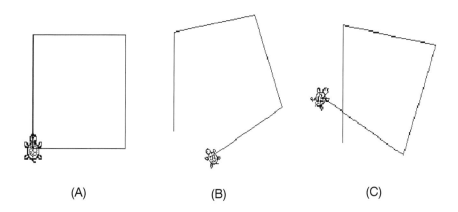

(A)   (B)   (C)

TABLE 1. Logo Tasks Worksheet

Shapes	Sides				Angles			
	a	b	c	d	A	B	C	D
1								
2								
3								
4								

the shape of the "given rectangle" with the four squares produced in Task One, and then correctly wrote the Logo code as "forward 100 right 90, forward 150 right 90, forward 100 right 90, forward 150 right 90" without using the "repeat" command (see Figure 2). Her successful completion of this task indicated that she could apply her new understanding of the concept of square to the construction of a rectangle. She produced another two rectangles and then used the worksheet without assistance to correctly summarize the characteristics of rectangle.

After we explained the requirements of Task Three, the participant performed well. Using Geometer's Sketchpad, she measured and then summed three angles of the triangle (see Figure 3). Then, as instructed, she dragged each angle to different positions. With each new position, she wrote down the changes of the angles, even though we did not provide her any worksheet for this task. Then she stopped to look at the changes of the measurements and sums of the different angles as they were displayed in the upper-right corner of the sketchpad screen. Finally, she concluded correctly that "the sum degree of three interior angles of a triangle is 180 degrees." However, although this child was able to apply her concrete understanding of the measurement and sum of angles to arrive at an abstract conclusion about the properties of a triangle, she was not completely clear about the generality of her conclusion.

Task Four required the child to use her abstract understanding about the properties of a triangle to draw conclusions about the sum degree of angles in

FIGURE 2. Logo and AC process (shape to code).

repeat 2 [forward 100 right 90 forward 150 right 90]

FIGURE 3. Geometer's Sketchpad and CA process.

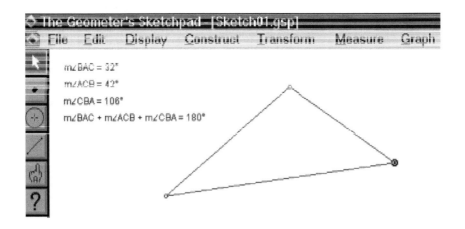

any quadrilateral shape. However, in this task, the child had difficulty deciding where to start because she could not identify and represent the problem. To assist her, we told her to construct any four-sided shape that was not a square or rectangle. She accomplished this task successfully, easily measuring and summing the angles. Then she dragged different angles of the shape and observed the changes in measurements (see Figure 4). Finally, she concluded that the sum of the angles was "360 degrees." However, as in Task Three, she still did not understand the generality of this conclusion–whether it would apply to "any" quadrilateral shape.

In summary, this child successfully completed each of the four tasks. However, she was not able to solve the tasks independently and needed guided assistance for each task. Because she learned best with concrete visual examples and verbal instructions, we placed her thinking at the concrete level. However, when provided with direct instructions and examples, she was able to complete all four tasks, an indication that her thinking/learning processes were capable of advancing to the abstract level via the process of concrete-abstract learning. With this particular child, therefore, it will be

FIGURE 4. Geometer's Sketchpad and AC process.

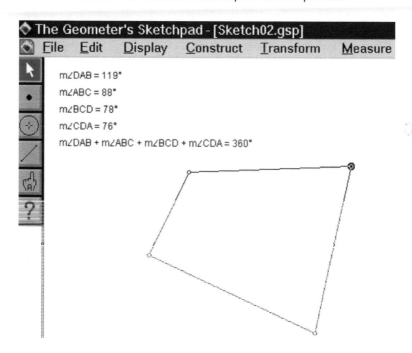

necessary for the teacher to lead her through the thinking/learning processes and to provide plenty of opportunities to practice these activities in order to facilitate movement to the abstract level.

Cases Two and Three

Cases Two and Three were observed at different locations and times. However, because of the similarities in performance on all four tasks, we have summarized them together. In Case Two, the participant was a ten-year-old girl; in Case Three, a ten-year-old boy. Both children were fourth-grade public school students. Their teachers and parents reported that their classroom learning performance and achievement were average.

Both children had previous computer experience with word processing, graphing, and games. Both of them also understood the instructions for using PCLogo and Geometer's Sketchpad. Also, after a quick demonstration of both technology tools, they followed the written instructions and quickly learned the skills required to perform the four tasks. Therefore, we concluded from the training segment that both of these children could learn well while working independently and could thus complete the four tasks using written instructions only.

In Task One, both children produced the first square shape correctly. In the next step, however, neither child followed the task requirements to replace the side-length numbers. One child (Participant Two) also replaced angle 90 with angle 80 and other angles of different degrees (see Figure 1-B). The other child (Participant Three) replaced the angle 90 with angle 120 and other angles of different degrees (see Figure 1-C). Both children also used various numbers after the repeat command, and then became excited about the strange shapes they created. Eventually, both children performed the task correctly, using various side lengths to create different sized squares. They printed out all of the shapes, compared them, and correctly summarized the definition of "square."

In Task Two, both children learned to use the "repeat" command (repeat 2 [forward 100 right 90 forward 150 right 90]) to create the rectangle shape (see Figure 2). Participant Three wrote the correct code on his first try. Participant Two first used "repeat 4 [forward 100 right 90 forward 150 right 90]" then changed to the "repeat 2" command. As in Task One, both children replaced the side lengths and changed the numbers of the angles just to see what shapes they could create. However, both were able to correctly summarize the characteristics of a rectangle.

In Task Three, both children followed the task requirements by measuring and summing the three interior angles of the triangle (see Figure 3). They arrived at the correct answer of 180 and understood that this was the sum of the angles of any triangle. They also measured the length of the sides of the triangles. Finally, when they dragged the angles of a triangle, they noted all of the subsequent changes in side lengths, degrees of individual angles, and the sum of the angles. However, the relationships among the side lengths remained a puzzle to them.

In Task Four, both children correctly followed all of the procedures. They first drew a four-sided shape, and then measured and summed the four angles (see Figure 4). Finally, both children were able to correctly summarize the sum degree of angles of any quadrilateral. Participant Three went even further, however, and measured the sum of the five interior angles of a pentagon and arrived at the correct answer of 540.

In both of these cases, there was no obvious evidence to suggest participants' initial thinking levels. However, in solving the four tasks, both performed the concrete-abstract and abstract-concrete tasks without requiring additional instructions. Both also demonstrated creativity in experimenting with different side lengths and angles. Thus, the CA and AC learning processes appeared to be used effectively in these two cases. The required tasks led them to the correct solutions as well as to other deductive solutions.

Case Four

Case Four was an 11-year-old boy, who was in the fifth grade and demonstrated average classroom learning and achievement. He was interested in, but not very good at, using a computer and had some word-processing and Web-search experience. During the training process, we provided him with written instructions for using PCLogo and Geometer's Sketchpad to perform the steps in each of the four tasks. However, he did not follow the exact steps in our instruction materials but experimented with the two programs and discovered on his own how to use them. Because he was able to learn independently of the concrete instructions, it is possible that his basic thinking skills were at the abstract level, although his geometric thinking level could not be determined.

In solving the first three tasks, this child easily followed the task requirements and obtained the correct answers, which were similar to those depicted in Figures 1-A, 2, and 3. In Task Four, the child wrote down the correct answer of 360, even before he began the task. He calculated the sum of the interior angles of a quadrilateral shape according to the principles he had learned in completing the previous tasks. We told him that the shape was called a "quadrilateral" and then asked him if his answer would apply to any quadrilateral shape. After experimenting with various quadrilaterals, he could correctly summarize the sum degrees of the angles and generalized his conclusions to shapes other than quadrilaterals.

The child in Case Four successfully performed the concrete-abstract and abstract-concrete thinking/learning tasks. His performance was different from that of the other participants, however, because he demonstrated both inductive and deductive reasoning. From specific situations (square and rectangle), he summarized the general rules (the sum of interior angles), and then used other specific situations (quadrilateral shapes created with Geometer's Sketchpad) to prove the general rule.

THE LEARNING MODEL

As a result of our observations of children's performances as they reasoned about geometric concepts, we have developed a learning model for teaching children about geometry. The model incorporates within it a description of children's reasoning processes that facilitate geometric thinking. Development of the model involved two steps:

1. We identified the reasoning processes required for geometric thinking and the technology tools that could effectively stimulate these processes. We thus assumed the following:
   a. the existance of CA and AC thinking processes;
   b. that CA and AC processes are required for summarizing and applying geometric concepts;
   c. that CA and AC processes can be stimulated through use of PCLogo and Geometer's Sketchpad; and
   d. that we could design learning tasks to stimulate the CA and AC thinking processes.

2. We concluded that, if children could successfully perform the learning tasks, then we could assume the following:
   a. the children effectively used CA and AC thinking processes, and
   b. PCLogo and Geometer's Sketchpad are useful tools for stimulating CA and AC thinking processes.

Keeping these assumptions in mind, we observed that in three of the cases, the children used CA and AC thinking to successfully perform the four tasks and arrive at correct solutions to the geometry problems. Only in one of the cases, Case One, did the child require concrete instructions before she could initiate and complete the tasks. However, even this child eventually used CA and AC thinking processes to correctly solve the problems. Therefore, the children's thinking processes and performance met the assumptions of the learning model, which we have conceptualized as the C-A-C Model depicted in Table 2.

In Table 2, the C-A-C pattern stands for *Concrete-Abstract-Concrete*, which represents the combination of the CA and AC processes. The basic assumption underlying the C-A-C model is that in geometric learning, children initially are presented with concrete tasks involving construction of geometric shapes. Completion of these tasks then leads children to reason abstractly about the geometric concept, or principle, that relates to the properties of the shape or figure. Finally, once they understand the underlying geometric concepts, children apply their abstract thinking processes to solve new but related concrete tasks. The four processes–CA and AC with PCLogo (produce shape with given code and write code to produce a given shape), CA and AC with Geometer's Sketchpad (measure and sum the angles of a given shape or object and draw conclusions about the geometric principle)–were integrated into the four tasks, which provided children with the practical experiences of using programming and exploratory software to stimulate geometric thinking.

The previous thinking/learning model described by Liu and Cummings (1997) was derived from traditional theories of schema construction and information processing. In the present study, we have identified and described a C-A-C learning model that is based on our own observations of children actually performing procedures designed to stimulate CA and AC thinking/learning processes. Hopefully, the descriptions of these tasks and the specific procedures required to complete them will provide mathematics teachers with practical strategies for using PCLogo and Geometer's Sketchpad in a classroom setting to stimulate learning about geometry.

CONCLUSIONS

A number of conclusions may be drawn from this case study. First, PCLogo and Geometer's Sketchpad were useful tools to stimulate children's geometric thinking by providing initial concrete experiences necessary for eliciting thinking about abstract geometric concepts. Although Geometer's Sketchpad was designed for high school students, it appears to be an effective tool for use by children in elementary grades if the learning tasks are effec-

TABLE 2. The C-A-C Learning Model

Pattern	C-A-C	
Processes	C-A	A-C
PCLogo	Code–Shape	Shape–Code
Geometer's Sketchpad	Shape–Measurement & Calculation	Construct procedures– Measurement & Calculation

tively designed, as they were in these four cases. These tasks now should be the focus of experimental studies that investigate the effectiveness of the model for teaching geometric concepts to children.

Second, the C-A-C model of learning appears to be useful for teaching geometry because its application encourages children to think, to explore, and to use different problem-solving methods regardless of their achievement levels. All of the children in this study were average in school achievement and had somewhat different learning styles. The model also would be appropriate for use with children whose academic achievement is above, or much above, average. Furthermore, it is possible that, with guided assistance and concrete use of the technology tools, children with below-average achievement levels might learn abstract principles of geometry through application of the C-A-C model. The model also accounts for different learning styles, such as independent exploration or learning with or without instructions. Finally, the model is useful because its application is congruent with a constructivist approach to teaching mathematics, which emphasizes critical thinking and problem solving rather than rote learning and memorization.

Third, the processes involved in advancing thinking from one level (CA) to another level (AC) occur as the child engages in repetitive practice of concrete skills through use of the technology tools. With enough practice, thinking processes undergo structural reorganization, producing a *qualitative* change in thinking that allows the child to think about geometry principles in a more complex and abstract way. For example, a child initially may use Logo commands experimentally, quickly learning how to draw sides of different lengths but failing to understand how to assign angle numbers to produce geometric shapes. Through continued experimentation, however, the child may accidentally produce a closed geometric figure (such as a triangle). This new discovery leads the child to notice the angle degrees as well as the numbers for the side lengths. Eventually, with enough practice, the child constructs an abstract understanding about the relationship of angles to geometric shapes, which represents a qualitative change in the child's thinking processes. These changes were observed to occur in the thinking processes of the children in the four cases in this study. It is our view that the goal of teaching always should be to lead children to construct these qualitative changes in thinking. Therefore, it is important that future studies examine: (a) how thinking advances from the concrete to the abstract level; (b) exactly how much repetitive practice with activities at the concrete level is required to stimulate qualitative changes in thinking; and (c) what other undiscovered factors may be responsible for construction of qualitative changes in thinking.

Fourth, previous models of cognition, such as those of van Hiele and Piaget, only describe the characteristics and applications of thinking at a particular level or stage (concrete or abstract). These models do not describe the thinking *processes* that advance thinking from concrete to abstract. However, understanding these processes may be more important than understanding the characteristics of a particular level or state of thinking. As a result of this study, we have developed a model of learning that describes the specific processes that advance thinking, and we have designed learning tasks that can be used to stimulate these processes. The tasks are useful for accomplishing two goals: (a) if the child's thinking is at the concrete level, the tasks will stimulate thinking to advance to the abstract level; and (b) if the child's thinking is already at the abstract level, the tasks will provide the practice necessary to prepare the child for another qualitative change in thinking.

In summary, findings of this study should provide mathematics teachers with practical suggestions for meeting NCTM standards that emphasize the integration of innovative technologies with classroom instruction. The C-A-C learning model is an effective model for explaining the processes that move thinking from the concrete to the abstract level, and the learning tasks described in this study can be incorporated into the elementary school mathematics curriculum. More generally, the idea of designing instructional tasks that stimulate qualitative changes in thinking has important implications for teaching and learning at all levels of instruction and with all age groups.

REFERENCES

Abramovich, S., & Nabors, W. (1997). Spreadsheets as generators of new meanings in middle school algebra. *Computers in the Schools, 13*(1-2), 13-25.

Clements, D. H., & Meredith, J. S. (1993). Research on Logo: Effects and efficacy. *Journal of Computing in Childhood Education, 4*(4), 263-290.

Davis, P. J., & Hersh, R. (1981). *The mathematical experience.* Boston, MA: Houghton Mifflin.

Denzin, N. K., & Lincoln, Y. S. (1994). *Handbook of qualitative research.* London: DAGE Publications.

Fuys, D., Geddes, D., & Tischler, R. (1988). The van Hiele model of thinking in geometry among adolescents. *Journal for Research in Mathematics Education Monograph Series, No. 3.* Reston, VA: National Council of Teachers of Mathematics.

Geddes, D. (1992). *Geometry in the middle grades: Curriculum and evaluation standards for school mathematics.* Reston, VA: National Council of Teachers of Mathematics. (ERIC Document Reproduction Service No. ED 351 200).

Geer, C. (1999). Geometry and geometric thinking. *Teaching Children Mathematics, 5*(6), 307-309.

Harvard Associates. (1994). *PC Logo for Windows.* Cambridge, MA: Author.

Hayes, J. R. (1989). *The complete problem solver* (2nd ed.). Hillsdale, NJ: Erlbaum.

Key Curriculum. (1999). *Teaching geometry with the Geometer's Sketchpad.* Emeryville, CA: Author.

Liu, L. (1999, March 1-4). *Technology and geometric concept learning.* Paper presented at the International Conference on Mathematics/Science Education & Technology, San Antonio, TX.

Liu, L., & Cummings, R. (1997). Logo and geometric thinking: Concrete-abstract thinking and abstract-concrete thinking. *Computers in the Schools, 14*(1-2), 95-110.

National Council of Teachers of Mathematics. (1989). *Curriculum and evaluation standards for school mathematics.* Reston, VA: Author.

National Council of Teachers of Mathematics. (1991). *Professional standards for schools for teaching mathematics.* Reston, VA: Author.

National Council of Teachers of Mathematics. (1999). Shaping the standards: Geometry and geometric thinking. *Teaching Children Mathematics, 5*(6), 358-360.

Papert, S. (1980). *Mindstorms: Children, computers, and powerful ideas.* New York: Basic Books.

Piaget, J. (1936). *The origins of intelligence in children.* London: International Universities Press.

Piaget, J. (1971). *Genetic epistemology.* New York: Basic Books.

Pokay, P. A., & Tayeh, C. (1997). Integrating technology in a geometry classroom: Issues for teaching. *Computers in the Schools, 13*(1-2), 117-123.

Swan, K. (1993). Domain knowledge, cognitive styles, and problem solving: A qualitative study of student approaches to Logo programming. *Journal of Computing in Childhood Education, 4*(2), 153-182.

Vacc, N. N., & Bright, G. W. (1999). Elementary preservice teachers' changing beliefs and instructional use of children's mathematical thinking. *Journal for Research in Mathematics Education, 30*(1), 89-101.

van Hiele, P. M. (1986). *Structure and insight.* Orlando: Academic Press.

van Hiele, P. M. (1997). *Structure.* Zutphen, Netherlands: Thieme.

van Hiele, P. M. (1999). Developing geometric thinking through activities that begin with play. *Teaching Children Mathematics, 5*(6), 310-317.

Weaver, C. L. (1991). *Young children learn geometric concepts using Logo with a screen turtle and a floor turtle.* (ERIC Document Reproduction Service No. ED 329 430).

Michael T. Battista

Shape Makers:
A Computer Environment That Engenders Students' Construction of Geometric Ideas and Reasoning

SUMMARY. Professional standards for school mathematics recommend that dynamic geometry programs such as the Geometer's Sketchpad can and should be used to enhance student learning of geometry. This article illustrates how a geometry computer microworld containing screen manipulable, dynamically transformable shape-making objects can promote the development of powerful geometric reasoning. Also described is the overall context and factors that should guide mathematics educators' design and use of instructional technology. *[Article copies available for a fee from The Haworth Document Delivery Service: 1-800-342-9678. E-mail address: <getinfo@haworthpressinc.com> Website: <http://www.HaworthPress.com> © 2001 by The Haworth Press, Inc. All rights reserved.]*

KEYWORDS. Geometry, computers, mathematics education, microworld, learning, teaching

MICHAEL T. BATTISTA is Professor, Mathematics Education, Kent State University, 404 White Hall, Kent, OH 44242 (E-mail: mbattist@kent.edu).

Partial support for this work was provided by grant ESI 9050210 from the National Science Foundation. The opinions expressed, however, are those of the author and do not necessarily reflect the views of that foundation.

[Haworth co-indexing entry note]: "Shape Makers: A Computer Environment That Engenders Students' Construction of Geometric Ideas and Reasoning." Battista, Michael T. Co-published simultaneously in *Computers in the Schools* (The Haworth Press, Inc.) Vol. 17, No. 1/2, 2001, pp. 105-120; and: *Using Information Technology in Mathematics Education* (ed: D. James Tooke and Norma Henderson) The Haworth Press, Inc., 2001, pp. 105-120. Single or multiple copies of this article are available for a fee from The Haworth Document Delivery Service [1-800-342-9678, 9:00 a.m. - 5:00 p.m. (EST). E-mail address: getinfo@haworthpressinc.com].

Current professional standards for school mathematics recommend that students doing mathematics and teachers teaching mathematics utilize appropriate technology (NCTM, 1989, 1998). In particular, the latest version of the *Standards* published by the National Council of Teachers of Mathematics (NCTM) suggests that dynamic geometry programs such as the *Geometer's Sketchpad* can and should be used to enhance student learning of geometry (1998). This new document also recommends that mathematics instruction be appropriately grounded in research on teaching and learning.

In this article, I illustrate how a specially designed dynamic geometry computer microworld can enhance students' geometric thinking. But before discussing the microworld, I describe the overall context and factors that should guide mathematics educators' design and use of instructional technology. This description not only helps us understand how best to use the geometry microworld, but also helps us appreciate that integrating technology into the mathematics curriculum without guidance from appropriate underlying principles–say, to "modernize" the curriculum–is highly unlikely to enhance student mathematics learning.

OVERALL CONTEXT FOR THE USE
OF TECHNOLOGY IN MATHEMATICS EDUCATION

Types of Technology Use

There are three basic types of technology used in mathematics education, all three of which have the potential to enhance student learning (although, often, that potential goes unrealized).

General technological tools include technology whose development is not driven by the needs of mathematics or mathematics teaching. An example is Web-based communication. Changes in general technological tools can change mathematics instruction, as these tools provide educators new ways to interact with students.

Technological tools for doing mathematics include technology developed outside the field of education for the purpose of doing mathematics more easily and powerfully. Examples include hand-held calculators and computers implementing software such as programming languages, spreadsheets, symbolic algebra, statistical packages, and graphing programs. Changes in technological tools for doing mathematics *should* cause educators to change school curricula to reflect current best practices in applying and doing mathematics.

Technological tools for teaching mathematics include technology that has been developed with the specific intention of enhancing student mathematics learning. In this category, I place educational software packages and instructional computing microworlds, an example of the latter being the focus of this article.

Basic Assumptions in Designing Technological Tools for Teaching Mathematics

There are several basic assumptions that I make in designing and using computer technology in mathematics instruction. First, technological tools for teaching mathematics must be reasonably easy to use for students and teachers. The technological interface must quickly recede into the background so that the focus of attention is on doing mathematics, not operating the technology. Second, because use of technological tools is an important component of modern applications of mathematics, there are clear advantages when students' use of technological tools for teaching mathematics helps them learn to use technological tools for doing mathematics. Third, and most importantly, technological tools for teaching mathematics must support instruction that is consistent with current professional standards for teaching mathematics and with modern research on mathematics learning. I will briefly outline such standards and research in the next two sections.

MATHEMATICS INSTRUCTION

The focus in the classroom environments envisioned by the NCTM *Standards* is on inquiry, sense making, and problem solving (1989). In such classrooms, teachers provide students with numerous opportunities: to solve complex and interesting problems; to read, write, and discuss mathematics; and to formulate and test the validity of personally constructed mathematical ideas so that they can draw their own conclusions. Students use demonstrations, drawings, real-world objects, and computing devices–as well as formal mathematical and logical arguments–to convince themselves and their peers of the validity of their problem solutions. In such classrooms, students develop competency not only with appropriate computation skills but also with mathematical reasoning and application.

MATHEMATICS LEARNING

All current major scientific theories describing mathematics learning agree that mathematical ideas must be personally constructed by students as they intentionally try to make sense of situations, including, of course, communications from others (Battista, 1999; De Corte, Greer, & Verschaffel, 1996; Greeno, Collins, & Resnick, 1996; Hiebert & Carpenter, 1992; Schoenfeld,

1994; Steffe & Kieren, 1994; Romberg, 1992). In this "constructivist" view, there are several key learning mechanisms. First, *abstraction* is the critical mechanism that enables the mind to construct the mental entities that individuals use to conceptualize and reason about their "mathematical realities." Meaningful mathematics learning results from the abstractions students make as they reflect on and adapt their current cognitive structures to deal with new situations and *perturbations*–realizations that one's current way of thinking or operating does not work or produces unexpected results. Understanding mathematics, however, requires more than abstraction. It requires *reflection*, which is the conscious process of examining in imagination experiences, actions, or mental processes and considering their results or how they are composed.

According to the constructivist view of learning, to get students to construct increasingly powerful mathematical ideas, instruction must promote abstraction and reflection. It must (a) focus student attention and mental acts on those aspects of phenomena that are to be mathematized; (b) encourage students' reflection on the viability of developing mathematical conceptualizations, and (c) promote appropriate perturbations to incorrect or unsophisticated student theories.

Another essential learning mechanism is the construction of mental models, which are integrated sets of abstractions that are activated to interpret and reason about situations that one is dealing with in action or thought. Mental models are nonverbal experience-like mental versions of situations whose structure is isomorphic to the perceived structure of the situations they represent (Battista, 1994; Greeno, 1991; Johnson-Laird, 1983). Individuals reason about a situation by activating mental models that enable them to mentally simulate interactions within the situation so that they can reflect on possible scenarios and solutions to problems.

What emerges from a careful consideration of the theory outlined above is a picture of meaningful mathematics learning coming about as individuals recursively cycle through phases of action (physical and mental), abstraction, and reflection in a way that enables them to integrate related abstractions into ever more sophisticated mental models of phenomena. In fact, individuals' ability to understand and effectively use our culture's formal mathematical systems to make sense of their quantitative and spatial surroundings depends critically on their construction of elaborated sequences of mental models. Initial models in these sequences enable individuals to reason about physical manipulations of real-world objects. Later models permit them to reason with mental images of real-world objects. Finally, symbolic models enable them to reason by meaningfully manipulating mathematical symbols representing real-world quantities. Without this recursively developed sequence of mental models, individuals' learning about mathematical symbol systems is strictly syntactic, and their use of symbolic procedures is disconnected from real-world situations.

Geometry Particulars:
The van Hiele Levels

To be consistent with the constructivist view of learning, mathematics instruction not only must create a classroom culture of inquiry that encourages and supports students' personal construction of meaning, but it must also base the design of instructional tasks on appropriate scientific research. Such research deals with (a) how students construct meanings for specific mathematical topics, (b) the stages that students pass through in acquiring targeted mathematical concepts and procedures, and (c) the strategies and reasoning that students use to solve problems during each stage.

In geometry, the best research-based description of the development of student reasoning is known as the van Hiele theory (Clements & Battista, 1992). According to this theory, geometric thought begins at the holistic visual level in which students identify and operate on shapes and other geometric configurations according to their overall appearance. Geometric thought progresses next to the point at which students can describe, analyze, and characterize shapes by their mathematical properties. At the next level, students' geometric thinking becomes more abstract and relational as they see that one property can imply other properties, define classes of shapes, distinguish between necessary and sufficient conditions for, as well as hierarchically classify, shapes. At the fourth level, students can comprehend and create formal geometric proofs; at the fifth, students can compare axiomatic systems.

CHARACTERISTICS
OF FERTILE COMPUTER ENVIRONMENTS

Instructional computer environments include not only the computer tools made available to students, but also the instructional tasks in which students are to apply these tools. As stated earlier, instructional computer environments–like all instruction–should structure instruction in ways that are consistent with research on mathematics learning. Thus, the design of instructional computing environments in geometry should be based not only on the general constructivist theory of learning, but also on research dealing with how students construct knowledge for particular geometric ideas.

To be consistent with the general constructivist theory, computer environments should engender and support genuine student problem solving and inquiry. Within the overall context of solving carefully designed sequences of problems in a classroom culture of inquiry, students should be able to make and test conjectures not only involving their proposed solutions to instructionally presented problems but also involving their own personally evolving mathematical conceptualizations. In fact, in an ideal environment, students would be encouraged to explore not only questions presented by the teacher but also ideas the students themselves generate.

Computer environments should also engender students' development and use of appropriate mental models for dealing with physical, conceptual, and symbolic mathematical phenomena. The instructional goal should be to encourage and support students' construction of mental models that enable them to make sense of, analyze, and solve problems concerning such phenomena.

Finally, computer environments should focus students' attention on phenomena in ways that support reflection on and abstraction of the mental operations necessary for properly conceptualizing and reasoning about mathematizations of phenomena. Because mathematical conceptualizations and associated mental models result from reflection on and abstraction of one's own mental actions, computer environments must make those actions and their consequences more accessible to reflection. Students' attention should be focused on developing and refining personally meaningful mathematical theories that guide their mathematical actions and reasoning. One especially effective device for maintaining such a focus is for students to make predictions before acting, which encourages them to consider their actions in the context of their theories. Inconsistencies between their predictions and what actually happens provide a constant source of perturbations requiring accommodations that lead to increasingly sophisticated conceptions.

THE SHAPE MAKERS COMPUTER MICROWORLD

The primary source of mental models is our experience in dealing with the world, especially with physical objects (Johnson-Laird, 1983). To appreciate how a mental model for a parallelogram might be derived from real-world manipulation, imagine four straight rods connected at their endpoints in a way that permits freedom of movement at the connections–a movable quadrilateral is formed (see Figure 1). Imagine now that the opposite rods are the same length. No matter how we move this physical apparatus, it always forms a parallelogram.

As we manipulate this "parallelogram maker," we not only see how its shape changes, but we also feel, and with proper reflection can abstract, the physical constraints that have been built into it–opposite sides equal and parallel. We see and feel how one parallelogram continuously changes into others. The visual and kinesthetic experiences that we abstract from our actions with this apparatus, along with our reflections on those actions, can be integrated to form a mental model that can be used in reasoning about parallelograms.

The Shape Makers computer microworld provides students with screen manipulable shape-making objects similar to, but more versatile than, the physical parallelogram maker described (Battista, 1998). In it, each type of common quadrilateral and triangle has a "Shape Maker," a Geometer's Sketchpad construction that can be dynamically transformed in various ways, but only to produce different shapes in that type. For instance, the computer Parallelogram Maker can be used to make any desired parallelogram that fits on the computer screen, no matter what its shape, size, or orientation–but only parallelograms. It is manipulated by using the mouse to drag its *control points*–small circles that appear at its vertices (see Figure 2).

The Shape Makers microworld was designed to promote in students the development of mental models that they can use for reasoning about geometric shapes. Through a sequence of carefully designed instructional activities with the Shape Makers, students are encouraged to pass through the first three van Hiele levels–from holistic visual thinking, to property-based description and analysis, and into the abstract relational thought that enables students to utilize logical relationships between properties and classes of shapes (Battista 1998). In initial activities, students use the various Shape

FIGURE 1. Movable parallelogram.

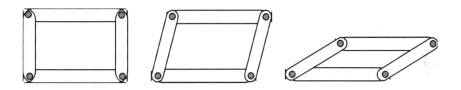

FIGURE 2. Parallelogram Maker in Geometer's Sketchpad.

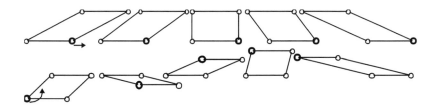

Makers to make their own pictures, then to duplicate given pictures. These activities encourage students to become familiar with the movement possibilities of the Shape Makers viewed as holistic entities. Students are then involved in activities that require more careful analysis of shapes; they are guided to find and describe properties of shapes. Unmeasured Shape Makers are replaced by Measured Shape Makers that display measures of angles and side lengths that are instantaneously updated when the Shape Makers are manipulated. Finally, students focus on classification issues as they compare the sets of shapes that can be made by each Shape Maker.

EXAMINING STUDENTS' THINKING
WHILE USING SHAPE MAKERS

Several episodes illustrate the types of thinking exhibited by students working in the Shape Makers microworld. All examples occurred in fifth-grade classrooms taught by a teacher who was skilled in creating a classroom culture of inquiry consistent with that promoted by the Shape Makers instructional unit.

Episode 1. In one of his initial activities with the Shape Makers, after MI tried to make a non-square rectangle with the Square Makers, he concluded that it couldn't be done:

> MI: *The Square [Maker] would only get bigger and twist around–so it can't make a rectangle.*

MI had abstracted several movement regularities of the Square Maker (it could be made bigger and rotated). However, he did not conceptualize these regularities precisely, as geometric properties, but thought about them vaguely and holistically, as is typical of students at van Hiele level 1. His statement that the Square Maker would only get bigger did not explicitly specify that it would remain a square (although it is likely that he recognized this), nor did he seem to think explicitly about side-length relationships (even though such considerations are implicit in his statement). MI had not yet developed a *conceptual system* that enabled him to precisely describe and completely understand the relevant spatial relationships he was observing.

Episode 2. Three students were investigating the Square Maker.

> MT: *I think maybe you could have made a rectangle.*

> JD: *No; because when you change one side, they all change.*

> ER: *All the sides are equal.*

MT, JD, and ER abstracted different things from their Shape Maker manipulations. MT noticed the visual similarity between squares and rectangles, causing him to conjecture that the Square Maker could make a rectangle. JD abstracted a movement regularity–when one side changes length, all sides change (thus, he couldn't get the sides to be different lengths, which he thought was necessary for a rectangle). So JD attended to a notion similar to that of MI, but JD explicitly used the idea of side length to conceptualize the situation more precisely (but still not with complete precision). Only ER conceptualized the movement regularity with complete precision by expressing it in terms of a traditional mathematical property. Thus, only ER's response was consistent with van Hiele level 2 reasoning.

Episode 3. Three students were considering whether the Parallelogram Maker could be used to make the trapezoidal target figure (Figure 3).

Their knowledge of the Parallelogram Maker was insufficient to predict that this was impossible. However, as they manipulated the Parallelogram Maker, one of the students discovered something that enabled her to solve the problem.

ST: *No, it won't work [pointing to the non-horizontal sides in the Parallelogram Maker]. See this one and this one stay the same, you know, together. If you push this one [a non-horizontal side] out, this one [the opposite side] goes out . . . This side moves along with this side.* (see Figure 4)

As ST manipulated the Parallelogram Maker in her attempts to make the target figure, she detected a pattern or regularity in its movement. As she abstracted this movement pattern, incorporated it into her mental model for

FIGURE 3. Target figure.

FIGURE 4. Parallelogram Maker.

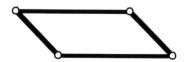

the Parallelogram Maker, and described it in terms of parts of the Shape Maker, she was able to infer that the target figure was impossible to make. By using the well-developed mental operations she had available for reasoning about physical objects, images, and motion, ST had formed a visual-kinesthetic mental model of the Parallelogram Maker that could, with further elaboration, serve as the basis for conceptualizing the formal mathematical property "in a parallelogram, opposites sides are parallel and congruent." Thus, this experience with the Shape Makers helped ST move toward level 2 thinking.

In summary, in all three of these episodes, we see students struggling to construct ways to conceptualize the spatial/geometric phenomena they are investigating. They make progress in their conceptualizations–moving toward property-based thinking–as they reflect on and analyze their manipulations of the Shape Makers.

Episode 4. NL is using the seven quadrilateral Shape Makers to make the design shown in Figure 5. A researcher is observing and asking questions as NL tries to make shape C with the Rhombus Maker.

NL: *The Rhombus Maker on [shape] B. It doesn't work. I think I might have to change the Rhombus Maker to [shape] C.*

Res: *Why C?*

NL: *The Rhombus Maker is like leaning to the right. On B, the shape is leaning to the left. I couldn't get the Rhombus Maker to lean to the left, and C leans to the right so I'm going to try it. [After her initial attempts to get the Rhombus Maker to fit exactly on shape C] I don't think that is going to work.*

FIGURE 5. Design to be made with Shape Makers.

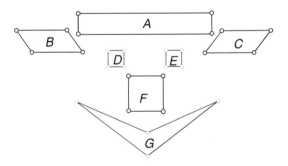

Res: *Why are you thinking that?*

NL: *When I try to fit it on the shape, and I try to make it bigger or smaller, the whole thing moves. It will never get exactly the right size. [Manipulating the Rhombus Maker] Let's see if I can make the square with this. Here's a square. I guess it could maybe be a square. But I'm not sure if this is exactly a square. It's sort of leaning. The lines are a little diagonal. [Continuing to manipulate the Rhombus Maker] Yeah, I think this is a square maybe.*

Res: *When you tried to fit it on C, did you notice anything about shape C, or the Rhombus Maker?*

NL: *The Rhombus [Maker] could make the same shape pretty much, but if you tried to make it small enough to fit on C, it would make the whole thing smaller or it would move the shape down. And when you tried to move it up to make it smaller, it would move the whole shape up.*

Res: *You said the Rhombus Maker could make the same shape as shape C, what do you mean by that?*

NL: *It could make the same shape. It could make this shape, the one with 2 diagonal sides and 2 straight sides that are parallel. It could have been almost that shape and it got so close I thought it was that shape.*

Res: *So that [the Rhombus Maker] is the same shape as that [shape C]?*

NL: *[Continuing to manipulate the Rhombus Maker] Oh, I see why it didn't work, because the 4 sides are even and this [shape C] is more of a rectangle.*

Res: *How did you just come to that?*

NL: *All you can do is just move it from side to side and up. But you can't get it to make a rectangle. When you move it this way it is a square and you can't move it up to make a rectangle. And when you move this, it just gets a bigger square.*

Res: *So what made you just notice that?*

NL: *Well, I was just thinking about it. If it [the Rhombus Maker] was the same shape, then there is no reason it couldn't fit in to C. But I saw when I was playing with it to see how you could move it and things like that, that whenever I made it bigger or smaller, it was always like a square, but sometimes it would be leaning up, but the sides are always equal.*

This episode clearly shows how a student's manipulation of a Shape Maker and resultant reflection on that manipulation can enable the student to move from thinking holistically to thinking about interrelationships between a shape's parts, that is, about its mathematical properties. Indeed, NL began the episode thinking about the Rhombus Maker and shapes holistically, saying that she was trying to make the Rhombus Maker "lean to the right," and get "bigger or smaller," and that "the whole thing moves."

The fact that NL could not make the non-equilateral parallelogram with the Rhombus Maker evoked a perturbation and caused her to reevaluate her mental model of the Rhombus Maker. Originally, because her model was not constrained by the property "all sides equal," her mental simulations of changing the shape of the Rhombus Maker included transforming it into non-equilateral parallelograms. Her subsequent attempts to make a non-equilateral parallelogram with the actual Rhombus Maker tested her model, showing her that it was not viable. As she continued to analyze why the Rhombus Maker would not make the parallelogram–why it would not elongate–her attention shifted to its side lengths. This new focus of attention enabled her to abstract the regularity that all the Rhombus Makers were the same length; that is, they were congruent. As she incorporated this abstraction into her mental model for the Rhombus Maker, she was able to infer that the Rhombus Maker could not make shape C.

It is highly likely that NL's conclusions about the Rhombus Maker came about partly because she had previously made a square with it. She viewed the Rhombus Maker as a transformed square, and we know from previous episodes that NL had already concluded that squares have all sides the same length. Using the Rhombus Maker took advantage of her and other students' natural proclivity to reason about shapes by mentally transforming them in various ways (Battista, 1994).

This example illustrates that students move toward property-based conceptions of shapes (and, therefore, to more sophisticated levels of geometric thinking) because of the inherent power these conceptions give to their analyses of spatial phenomena. In the current situation, NL developed a property-based conception of the Rhombus Maker because it enabled her to understand why the Rhombus Maker could not make shape C–something that truly puzzled her.

Episode 5. Shape Makers also provide students with "concrete" ways to think about *classes* of shapes, a topic that middle school students have considerable difficulty with. The Rectangle Maker, for example, can be used to think about the class of rectangles because the properties that constrain its movement are exactly the properties possessed by all rectangles. As the episode below illustrates, this Shape Maker to shape-class correspondence can enable students to interrelate classes hierarchically, a characteristic of van Hiele level 3.

BE: *A square is a rectangle, but a rectangle is not a square.*

MA: *I agree. The Rectangle Maker can make a square, but the Square Maker cannot make all rectangles.*

SO: *Every shape made by the Square Maker can be made by the Rectangle Maker because a square is a rectangle.*

MA's and SO's statements show how they were using their knowledge of the Shape Makers to make sense of and justify BE's claim that squares are rectangles. Because MA and SO had properly connected the Rectangle and Square Makers with the classes of shapes made by these Shape Makers, these two students could reflect on their mental models of the Shape Makers and draw conclusions about properties of, and interrelationships between, classes of shapes.

Episode 6. BT and LR were predicting whether the Measured Parallelogram Maker would make shapes 1-6 shown in Figure 6.

BT: *I think the Parallelogram Maker can make it [shape 1] 'cause these lines are parallel. Right and left lines are parallel, and top and bottom, opposite lines are parallel.*

LR: *Oh my gosh! Look, look, look! The angles are different [pointing to the angle measures on the screen]; angle A is 55.8 degrees and angle D is 124.2 degrees. And that's [pointing to angle C on the Parallelogram Maker] the same as angle A; A matches up with C.*

BT: *These two are opposite [pointing to angles A and C on the Parallelogram Maker]. And these two [pointing to angles B and D] have to have the same angles too.*

LR: *[Manipulating the Parallelogram Maker–so the girls could see that opposite angle measures remained the same as they changed its shape] These 2 always have to follow each other! [Excitedly] Write it down, write it down!*

FIGURE 6. Measured Parallelogram Maker activity.

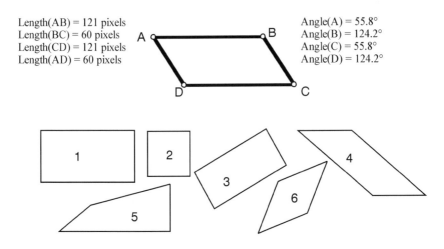

Length(AB) = 121 pixels
Length(BC) = 60 pixels
Length(CD) = 121 pixels
Length(AD) = 60 pixels

Angle(A) = 55.8°
Angle(B) = 124.2°
Angle(C) = 55.8°
Angle(D) = 124.2°

BT and LR noticed and explicitly attended to the fact that opposite angles in a parallelogram are congruent, even though the instructional task did not require attention to this concept. Both the classroom culture and the Shape Makers instructional environment invited and provided opportunities for students to make serendipitous property-based discoveries. Moreover, because the Shape Makers microworld is inherently interesting to students, and because it permits students to pursue their own ideas, not just those targeted by the teacher or curriculum, the majority of students actually get excited about learning.

CONCLUSIONS

The instructional use of the Shape Makers microworld described above is consistent with professional standards for teaching mathematics and with current research on student learning of mathematics in general, and geometry in particular. In the spirit of problem solving and inquiry, students working with the Shape Makers develop rich mental models that enable them to reason in increasingly sophisticated ways about important types of shapes. They gradually come to know the geometric properties of shapes, not as empty verbal statements to be memorized, but as powerful conceptualizations that enable students to sharpen their geometric analyses of spatial phenomena. In this way, students' work with the Shape Makers microworld helps them develop one of the most important goals of the NCTM *Standards*, "mathematical power," a term denoting students' ability "to explore, conjecture, and reason logically, as well as the ability to use a variety of mathematical methods effectively to solve nonroutine problems. This notion is based on the recognition of mathematics as more than a collection of concepts and skills to be mastered; it includes methods of investigating and reasoning" (1989, p. 5).

Thus, the Shape Makers microworld is an extremely powerful technological tool both for teaching a number of critical geometric ideas and for cultivating reasoning techniques that increase students' overall mathematical power. However, an added benefit of using the Shape Makers microworld is that it introduces students to an extremely powerful technological tool for *doing* mathematics–the Geometer's Sketchpad. Unlike ubiquitous computer drill-and-practice programs, using Shape Makers actually involves students in doing mathematics with technology in a way that anticipates their later use of dynamic computer graphing and visualization tools as sophisticated adults. (For instance, I used the Geometer's Sketchpad to model the planting of pine trees in my back yard. I wanted to analyze the effects of tree placement and spacing on the views from various windows in the house.) Thus, the Shape Makers microworld not only enhances students' learning of important mathematical ideas, it involves students in using a powerful technological tool that they can utilize throughout their mathematical careers.

REFERENCES

Battista, M. T. (February 1999). The mathematical miseducation of America's youth: Ignoring research and scientific study in education. *Phi Delta Kappan, 80*(6), 424-433.

Battista, M. T. (1994). On Greeno's environmental/model view of conceptual domains: A spatial/geometric perspective. *Journal for Research in Mathematics Education, 25*, 86-94.

Battista, M. T. (1998). Shape MakerS: Developing geometric reasoning with the Geometer's Sketchpad. Berkeley, CA:Key Curriculum Press.

Clements, D. H., & Battista, M. T. (1992). Geometry and spatial reasoning. In D. Grouws (Ed.), *Handbook of research on mathematics teaching and learning* (pp. 420-464). New York: NCTM/Macmillan.

De Corte, E., Greer, B., & Verschaffel, L. (1996). Mathematics teaching and learning. In D. C. Berliner & R. C. Calfee (Eds.), *Handbook of educational psychology* (pp. 491-549). New York: Simon & Schuster Macmillan.

Greeno, J. G. (1991). Number sense as situated knowing in a conceptual domain. *Journal for Research in Mathematics Education, 22,* 170-218.

Greeno, J. G., Collins, A. M., & Resnick, L. (1996). Cognition and learning. In D. C. Berliner & R. C. Calfee (Eds.), *Handbook of educational psychology* (pp. 15-46). New York: Simon & Schuster Macmillan.

Hiebert, J., & Carpenter, T. P. (1992). Learning and teaching with understanding. In D. A. Grouws (Ed.), *Handbook of research on mathematics teaching* (pp. 65-97). Reston, VA: NCTM/Macmillan.

Johnson-Laird, P. N. (1983). *Mental models: Towards a cognitive science of language, inference, and consciousness.* Cambridge, MA: Harvard University Press.

National Council of Teachers of Mathematics. (1989). *Curriculum and evaluation standards for school mathematics.* Reston, VA: Author.

National Council of Teachers of Mathematics. (1998). *Principles and standards for school mathematics: Discussion draft.* Reston, VA: Author.

Romberg, T. A. (1992 November). Further thoughts on the Standards: A reaction to Apple. *Journal for Research in Mathematics Education, 23*, 432-437.

Schoenfeld, A. C. (1994). What do we know about mathematics curricula? *Journal of Mathematical Behavior, 13*, 55-80.

Steffe, L. P., & Kieren, T. (1994). Radical constructivism and mathematics education. *Journal for Research in Mathematics Education, 25*(6), 711-733.

Robert D. Hannafin
Barry N. Scott

Teaching and Learning with Dynamic Geometry Programs in Student-Centered Learning Environments: A Mixed Method Inquiry

SUMMARY. The present study examined the attitudes and beliefs held by four middle school teachers and their 226 students about teaching and learning geometry using Geometer's Sketchpad in a student-centered environment. Findings indicate that teachers liked and valued Sketchpad as a tool but had conflicting opinions about the value of using it to support student-centered activities. Students were quite positive about controlling their own learning and their ability to interact dynamically with onscreen geometric shapes. Findings indicate that the task of changing teacher attitudes about the school learning environment may be more difficult than researchers and educational reformers have presumed. *[Article copies available for a fee from The Haworth Document Delivery Service: 1-800-342-9678. E-mail address: <getinfo@haworthpressinc. com> Website: <http://www.HaworthPress.com> © 2001 by The Haworth Press, Inc. All rights reserved.]*

ROBERT D. HANNAFIN is Professor, School of Education, College of William and Mary, 232 Jones Hall, Williamsburg, VA 23187-8795 (E-mail: rdhann@facstaff.wm.edu). BARRY N. SCOTT is Doctoral Candidate, Instructional Technology, Auburn University, Auburn, AL 35016 (E-mail: scottbn@auburn.edu).

[Haworth co-indexing entry note]: "Teaching and Learning with Dynamic Geometry Programs in Student-Centered Learning Environments: A Mixed Method Inquiry." Hannafin, Robert D., and Barry N. Scott. Co-published simultaneously in *Computers in the Schools* (The Haworth Press, Inc.) Vol. 17, No. 1/2, 2001, pp. 121-141; and: *Using Information Technology in Mathematics Education* (ed: D. James Tooke and Norma Henderson) The Haworth Press, Inc., 2001, pp. 121-141. Single or multiple copies of this article are available for a fee from The Haworth Document Delivery Service [1-800-342-9678, 9:00 a.m. - 5:00 p.m. (EST). E-mail address: getinfo@haworthpressinc.com].

121

KEYWORDS. Dynamic geometry, Geometer's Sketchpad, mathematics, technology, student-centered learning environments, teacher's role

In recent years, many educational reformers have advocated using computer technology to create more learner-centered, open-ended learning environments (OELEs) where learners are provided with varying amounts of help and support in deciding what they need to learn and what resources they need to learn it (e.g., Cognition and Technology Group at Vanderbilt [CTGV], 1992; Hannafin, 1992; Land & Hannafin, 1996). Proponents of OELEs believe that, by identifying goals and constructing meanings, learners become active managers, rather than passive receptacles, of information. According to Land and Hannafin (1996), one characteristic of an OELE is that it provides learners with opportunities to engage the environment in ways that support their unique needs and intentions for making sense of the world. OELEs do not provide predetermined content that follows structured learning objectives, rather OELEs assist learners in defining goals and generating and revising appropriate learning strategies. Hannafin (1996) described OELEs as:

> systems designed to support the unique search and understanding needs of individuals. That is, they are not designed to teach particular content, to particular levels, for particular purposes; they are designed to support learners' attempts to understand for their own purposes. In effect, OELEs impose no particular pedagogical strategy or instructional sequence but guide learners in invoking their own strategies and generating their own learning sequences. (Online)

In contrast with the assumptions underlying OELEs, most traditional instructional programs are grounded in an objectivist learning theory, which holds that knowledge exists outside of the mind and must be transmitted to the learner by some means, primarily a teacher, for learning to occur (Lakoff, 1987). The authenticity of learning outcomes in the objectivist learning environment, which has been dominant in most public schools, has been questioned by advocates of school reform (Means, 1994).

In comparison, Rieber (1992) used the term *instructivism* to describe an approach to designing instruction that falls somewhere between learner-centered constructivism, the philosophy generally consistent with the creation of OELEs, and content-centered objectivism. An instructional program designed from an instructivist approach would allow for externally imposed goals but would try to accommodate the unique cognitive strategies of individual learners. One application of instructivism has been the design of microworlds, which afford learning activities that match learners' current needs and prior knowledge but which also simplify and limit the type and amount of content available to the learner.

Whether students are afforded great freedom in managing their own learning, as in the case of truly open-ended environments, or are somewhat restricted in the paths they may take, as in microworlds, the need to provide instructional support and guidance is, of course, critical. Notwithstanding the strides in the development of intelligent tutoring and artificial intelligence systems (Bielawski & Lewand, 1991; Elsom-Cook, 1990; Schank & Cleary, 1995), that support is most often in the form of what Meyer (1992) called a "knowledgeable other" (p. 42). The skillful coaching of the classroom teacher is widely believed to be the key factor in predicting student success in less structured learning environments (Choi & Hannafin, 1995; Land & Hannafin, 1996; Young, 1993).

Teachers can provide instructional scaffolds that assist students in going from the unknown to the known. The gap in knowledge or skill acquisition being bridged through scaffolding is most often referred to as the *zone of proximal development* (ZPD) (Meyer, 1992; Palincsar & Brown, 1984; Vygotsky, 1978). The ZPD is the gap, or distance, between what a learner can achieve independently and what he can achieve with the help of a more knowledgeable peer (Vygotsky, 1978). The role of the teacher in guiding his or her student through a ZPD is not so much to provide verbal explanation of a task as it is to *participate* in the task alongside the student, guiding the student through the task and anticipating what assistance the student will need to accomplish it (Rogoff & Gardner, 1984). Teachers can provide scaffolding in the form of cueing, prompting, analogies, metaphors, questioning, elaborations, and modeling. The ultimate goal of scaffolding is to be able to fade and remove a scaffold as the student takes on the full responsibility of completing the task (Meyer, 1992). A typical teacher who is accustomed to providing group instruction–with each student working on the same task in the same manner–may have a difficult transition to the more demanding practice of instructional scaffolding in more open-ended environments.

The importance of the teacher's role in facilitating student inquiry in OELEs is unquestioned; teachers' ability or willingness to adapt their teaching style to facilitate such inquiry is, however, another matter. The Cognition and Technology Group at Vanderbilt (1992) found that teachers who used the *Jasper* videodisc series, a series that situated the learning of mathematics concepts in authentic, real-life contexts, tended to follow one of three instructional models: basics-first; structured problem solving; or guided generation. Teachers who used the basics-first model believed that students must first be armed with basic prerequisite facts before they can successfully solve complex problems. The second model allowed for the teacher to define a set of instructional objectives and to chart the best instructional path to accomplish those goals. The third model required the *students* to identify the problem as well as the strategy and information needed to solve it. The teacher's role under the guided generation model is that of facilitator and scaffold builder, helping students to think about the nature of the problem and how to go about solving it on their own. This model allows for the teacher to help students work through and correct misconceptions; whereas, the first two models tend to stress direct instruction and the avoidance of errors.

Many teachers may intuitively and/or conceptually see the value in these types of student-centered, teacher-guided instructional transactions, but it is often difficult to put them into practice (Scott, 1998). Barriers that may inhibit practice informed by one's own theories and beliefs include the culture and traditions of schooling (Tobin & Dawson, 1992), external pressures (Scott, 1998), lack of faith in students' abilities to take control of their own learning (Hannafin & Freeman, 1995), the inability to leave one's comfort level (Saye, 1997), and even student expectations about how learning should occur (Saye, 1997). Hannafin (1999) found, through three graduate courses, that attempts to change experienced teachers' attitudes about learning were unsuccessful. It is important to learn more about these barriers and how to overcome them since teachers are the ultimate gatekeepers for the kinds of instructional activities that occur in their classrooms.

Dynamic geometry programs, such as Geometer's Sketchpad (1993), are not OELEs in and of themselves; rather, they are tools, much the same as word processors and spreadsheets that can be used to create and support student-centered learning environments. They contain no information or instructional content. Tools available in dynamic geometry programs could be used just as easily by teachers with an objectivist philosophy of learning as they could be by teachers with a more constructivist view. Pea (1993) stressed the importance in the distinction between simply *providing* technological tools that facilitate student inquiry and how those tools are actually *used*: "These (technology) tools provide an *opportunity* to engage in higher-order thinking, but do not inherently enhance cognitive activity or skills" (pp. 41-42).

The designers of the Geometer's Sketchpad recommend that, to encourage student-directed inquiry, a series of structured activities and guiding questions be provided to guide students to the point where conjectures are possible. Such activities are designed to promote learning but do not regulate how the tool is used. Learners postulate and make conjectures as they "click and drag" geometric shapes. Learners are encouraged to work collaboratively to formulate theories and draw their own conclusions. While these kinds of designer-orchestrated activities are perhaps more consistent with Rieber's description of an instructivist environment, some (e.g., Goldenberg & Cuoco, 1996) assert that students can learn to think critically, become better problem solvers, and are better able to draw and transfer meaning when afforded these opportunities. For example, in a study using structured activities designed for the Sketchpad, Hannafin and Scott (1998) reported that, while high-ability students predictably outscored their low-ability counterparts on lower order factual recall posttest items, their scores on the higher order posttest items, although higher, were not significantly so.

Mathematics educators are intuitively attracted to dynamic geometry programs like Sketchpad, sensing that powerful learning outcomes are possible (Goldenberg & Cuoco, 1996). But while considerable research has been done involving other technologies like mathematics graphing (Dunham, 1993), few investigations so far have examined teachers' or students' attitudes about using dynamic geometry tools in environments less structured than their normal classroom. Examining teachers' attitudes may help identify areas of potential resistance to the shift in role to facilitator and inconsistencies between their theoretical view and actual practice. In addition, examining students' attitudes and ability to manage their own instruction under somewhat less structured environments would seem to have critical instructional development implications. This study permitted examination of the following research questions:

1. Does modeling (for teachers) facilitation skills in a student-centered learning environment, using a dynamic geometry program, influence teachers' beliefs about this type of learning environment and their students' abilities?
2. What are students' perceptions and reactions to learning geometry in less structured environments with dynamic geometry programs like the Sketchpad?

METHOD

Participants

Participants were four eighth-grade mathematics teachers in a middle-class city located near a major southeastern university. They comprised the entire math faculty at the host middle school. Mathematics achievement is historically low in the host school. The 1996 Stanford Achievement Test composite mathematics scores were at the 43rd percentile. Each teacher taught three eighth-grade classes. The total of 226 students (110 boys, 116 girls) participated in the study.

The project was attractive to the eighth-grade faculty for three reasons: (a) it provided them an opportunity to learn how to use Sketchpad in a non-threatening setting; (b) it gave them a break from their normally rigorous teaching schedule, yet covered the state-mandated geometry strand required for eighth grade; and (c) it represented a needed change of pace for their students. They agreed to allow the authors to develop instructional materials based on the state-mandated math curriculum to use in conjunction with the Geometer's Sketchpad program and to deliver the instructional unit in their regularly scheduled eighth-grade class meetings. Each class took one week to complete the program. The teachers essentially agreed to relinquish control of their classes for the duration of the study. Our role was to act as coaches or facilitators during the sessions and not to provide explicit answers or pre-scriptive directions on how to complete an activity. The teachers brought their classes to the lab and got them settled down to work, enforced discipline when necessary, and tried to learn to use the Sketchpad for future classes. We encouraged teachers to help students during the activities as long as they did not provide explicit answers or overly prescriptive directions. However, no training was provided to prepare them to make the transition from the tradi-tional lecture-based model to the more student-centered approach employed in the current program.

Brief biographies using pseudonyms of the participating teachers follow:

Ben, a retired military officer, is a white male in his late 40s who had been teaching for four years. He appeared to be very regimented and organized in his personal life, but his class management style was comparatively loose and surprisingly undisciplined. He seemed to have genuine concern for his students. He lamented the state of education today but did not blame the students. He bluntly acknowledged his own limitations and was open and willing to try new ideas to motivate and stimulate students.

Barb, a white female in her late 40s, had been teaching at the host school for about 15 years. She was born and raised in the host school district and attended the host school as a student. She talked about adopting new strategies and of integrating technology into her teaching but admitted she had not tried to do so yet.

Karen, a white female in her late 20s, had taught for four years. She seemed confident in her ability to use technology in her classroom, although she admittedly had not much opportunity. She also seemed open and anxious to learn how to use Sketchpad.

Kim, a white female in her late 50s, had 20 years' teaching experience and was the department chair. She was very organized. She rarely used the computer lab and did not use any other technology in her classes, although she acknowledged the importance of teaching and learning with technology.

The 226 students took part in the study during their normally scheduled mathematics periods. Two of the 12 participating classes were pre-algebra and were above grade level (based on their mathematics Stanford Achievement Test), two other classes were below grade level, and the remaining eight classes were considered to be near or at grade level in mathematics. The content covered in the instructional program was designed to satisfy a state-mandated geometry strand required for all eighth-grade mathematics classes. Mathematics achievement is historically low in the host school.

Materials

Advocates of OELEs stress not only the importance of the teacher's role as facilitator and diagnostician but also that sufficient structure be provided in OELEs to allow students of all ability levels to manage their learning (Hannafin, Land, & Oliver, 1999). In the present study, a booklet containing 16 structured activities covering geometric concepts was developed by the authors. These activities were designed to allow for student inquiry while guiding, prompting, and helping students to identify relationships and make conjectures. The booklet was intended to provide as much guidance as possible to allow students to work independently. The participating teachers, along with the authors, were available to clarify, guide, and assist students to work through the activities, but not to provide explicit answers. The activities progressed in difficulty from basic geometric concepts, such as measuring and classifying angles, lines, rays, and segments, to more advanced concepts dealing with formulas and relationships in circle geometry. The same booklet was used for all twelve classes.

The activities directed students to draw, measure, and manipulate segments, angles, circles, and other geometric shapes. Guiding questions accompanied each of the activities. The questions asked learners to describe relationships among the on-screen objects as a result of their manipulations. In the activity depicted in Figure 1, students "dragged" one of < CAB's (constructed and measured in the previous activity) sides until the measure of the angle was 90 degrees (the measure of the angle was displayed on screen and changed dynamically as the angle was altered). Students were then instructed to draw a segment (AD) from the angle's vertex to a point (D) inside the angle sides. Next students were directed to measure the newly formed angles CAD and DAB, and then to "drag" segment AD and observe the changes in the measure of all three angles. As students completed each of the activities, they advanced to the next activity on their own. At the beginning of each new period, they returned to the previous day's exit point.

FIGURE 1. Example of instructional activity.

Swap jobs with your partner. Continue using the sketch from the previous activity.

1. Change ∠BAC so that it is a right angle.

2. Draw a new line segment between the two angle sides. Look back to page 3 if you need a reminder.

3. How many angles do you see now? _____
   Your sketch should look similar to this one:

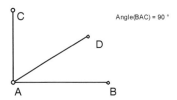

4. Measure the two small angles. (Look back to page 6 if you need a reminder.)

   Now, you should see three angle measures on your sketchpad. One for the large angle, and one for each of the two smaller angles.

5. Drag the endpoint D of the center line segment to the right or left (staying inside the larger angle).

   What happened to the measures of the three angles? _____

   Did the 90º angle change? _____

6. Add the two smaller angles together.

   What is the sum of the two angles? _____

   What can you say about the relationship between the two small angles and the larger angle? _____

7. Name two adjacent angles. _____

   The two angles formed inside the original right angle are adjacent and complementary. The two small angles you have in your sketch are adjacent and complementary no matter how you drag the middle segment's endpoint.

8. What do you think adjacent means? _____

   What do you think complementary means? _____

   To learn more about the terms *adjacent* and *complementary*, turn to page 218 in your textbook.

Procedures

Prior to the initial experimental session, students were informed by their regular teacher that their test grade at the end of the program (not included in the present report) would count toward their course grade. Students participated in all experimental sessions in a dedicated computer lab. The program was administered during regularly scheduled classes over a period of approximately four days.

Consistent with Sketchpad's designers' and others' (CTGV, 1992 & 1993; Hooper, 1992a, 1992b; Singhanayok & Hooper, 1998) recommendations to provide opportunities for student collaboration, students were assigned to work with a partner throughout the 16 activities. Based on Hooper's (1992b) recommendations on heterogeneous ability grouping, students were paired based on their year-to-date mathematics grades. The student with the highest mathematics grade was paired with the student with the lowest mathematics grade; the student with the second highest grade, with the student with the second lowest grade, and so on.

The participating teachers reported that students were accustomed to working in groups and were instructed to work and collaborate with their partners during the program. Some degree of student collaboration was ensured by performing interdependent tasks. For example, one member of each dyad was designated to record the team answers in the activities booklet (e.g., Figure 1) for an entire activity while the partner drew and manipulated the on-screen shapes and sketches. Students then traded jobs for the next activity, alternating roles throughout the entire program. Each dyad worked independently, and students were reminded to use their textbooks to look up definitions of new terms presented in the booklet. The booklet provided page number references to important terms found in the textbook, but no definitions were provided in the booklet. Students also received help from the researchers and the teachers in using the tools available in Sketchpad. Correct answers to the 16 activities were not provided during the activities. Thus, students were allowed to test their conjectures, even in cases where their reasoning appeared to be faulty. Correct answers were provided during a 20-minute comprehensive review provided by the authors at the end of the program. The review allowed students to compare their booklet answers with the correct answers for the first time. Students were allowed to complete the program at their own pace (but, in reality, slower students undoubtedly felt somewhat pressured to finish).

Instruments and Data Collection

Observations. The participating teachers were observed during all the sessions by the two authors. Two doctoral students also observed about half of the sessions. The authors also observed student interactions and recorded selected student comments.

Interviews. The doctoral students interviewed each teacher individually for about 40 minutes during the last class in the final week of the experimental sessions. An interview guide (Patton, 1990) was prepared prior to the interviews to ensure consistency in the basic areas covered. The interviews started with a brief discussion about the teacher's past and then proceeded to inquiries about classroom management, teaching philosophy, perceptions about the Sketchpad program, students' reactions, and computer use in the classroom. The interviewers followed the line of questioning that seemed interesting and relevant to the experience of the research and that might illuminate teachers' attitudes about the assumptions underlying student-centered learning.

Surveys. The four teachers completed an attitude survey following the program. The survey, displayed in Table 1, included nine Likert-type items, where teachers reported their agreement or disagreement on a scale from 1 (Strongly Agree) to 4 (Strongly Disagree), and seven open-ended items. An attitude survey was also administered to students immediately following the program. The student survey, included in Table 2, included eight Likert-type items and three open-ended items.

Design and Data Analysis

Teacher observations were primarily informal and were targeted to answer the specific research questions outlined above. Correlations among student attitudes were examined using Pearson Product Moment tests. Tests were performed at an alpha level of .05. The teacher interview data as well as student responses to open-ended survey items were analyzed and reduced into a smaller number of convergent themes. The multiple data sources and analysis methods were used to achieve triangulation (Patton, 1990), which provided a richer, more detailed picture of the phenomena under study and strengthened the validity and credibility of the findings.

RESULTS

The results are organized around the two stated research questions.

The First Research Question

1. *Does modeling facilitation skills in a student-centered learning environment using a dynamic geometry program influence teachers' beliefs about this type of learning environment and their students' abilities?*

In the teacher interviews, all four teachers described their normal teaching style as "almost entirely lecture mode," relying heavily on the textbook as the curriculum anchor. Interestingly, they were almost apologetic when describing their normal teaching styles and were quite supportive of creating more student-centered learning environments in the future. They were also excited about the opportunity to learn Sketchpad in a safe environment, where they could allow us to handle any technical problems and they did not have to be "on the spot" in front of their students. Their reliance on the lecture mode was not surprising, given the fact that the eighth-grade mathematics curriculum is demanding and the pressure to cover content often forces teachers, even those who would prefer to do otherwise, to sacrifice depth of coverage for breadth. In addition, anxiety brought on by the pressure to increase math scores on the year-end Stanford Achievement Test (SAT) was palpable. Karen, one of the eighth-grade teachers, observed that "everything is going to come down to this one test. How your school is performing is based on this one test. And so there's really been a push to make sure everything is covered." The screen-savers in the computer lab graphically compared the host school's unfavorable SAT percentile ranking to neighboring school districts as a constant reminder to students. Discipline and motivation were constant challenges to these teachers. The teacher survey mean scores, summarized in Table 1, indicated very favorable attitudes about the instructional program.

Despite the teachers' enthusiasm and good intentions prior to the program, only Karen and Ben, in our view, would be able to make a reasonably successful transition in teaching style. Karen had an easy rapport with her students, made full eye contact, and often would touch the shoulder of the student she coached. She was very active and "worked" the room but did not intervene unless students were struggling. She often tried to scaffold the instruction, asking her students probing questions enabling them to develop strategies to progress in the problem-solving activities. She learned to use the software very well and would be able to teach with Sketchpad on her own with no difficulty. In terms of her students, she was very encouraged, noting that both the high-ability and low-ability students benefited: "When I asked one of my better students who was working with a very low student who needs confidence how they were doing, they both replied 'Oh this is easy; this is fun!'" Another young man, whom Karen earlier described as "running at the mouth all the time" and "having difficulty sitting and listening to someone else in class," surprised Karen by staying on task and seemed to get a lot out of this experience because he was able to "do it himself." Karen also reported that she was very encouraged by how many students used the textbook to look up terms and answers. "It's been neat to watch them go through the book and try to figure stuff out, because they normally won't do it."

TABLE 1. Teacher Survey Items and Attitude Responses

Item	Teacher			
	Ben	Barb	Karen	Kim
1. Overall, I liked the program	SA	SA	SA	SA
2. My students seemed to try hard to do well in the program	SA	A	A	A
3. I felt uneasy about letting students control their own learning	SD	A	D	SD
4. I like activities where students have a chance to figure things out for themselves	SA	A	A	SA
5. Students learn better working with a partner	SA	A	D	A
6. The program made my students think a lot	A	A	SA	A
7. I think students need more help from the instructor to do well during the program	A	SA	A	D
8. A program like *Sketchpad* would be more effective with longer blocks of class time	D	SA	SD	D
9. A program like *Sketchpad* is better suited for higher-ability students	D	SA	SD	A
10. What did you like best about the program?				
11. What did you like least about the program?				
12. How was this program different from your regular classroom?				
13. Will you use *Sketchpad* in the future? If yes, how? If not, why not?				
14. Do you feel confident in your ability to structure activities that involve the *Sketchpad*? If no, why not?				
15. Based on your brief observations of the *Sketchpad*, how would you compare students' depth of learning using *Sketchpad* with that of regular instructional practices?				
16. How often do your lesson plans include computer use?				

Interestingly, however, despite Karen's encouraging reactions about her students, she did not agree philosophically with some of the program's underlying assumptions. She expressed that, although she honored our request not to provide direct instruction beforehand, she would have preferred to provide definitions of basic terms prior to the experimental sessions: "This is the one thing I would have liked–to present my kids with vocabulary and learn that before they start working with it [Sketchpad] so they would know what they were doing beforehand instead of just being thrown into it." Also, although Karen believed that all of her students benefited in some way, she believed that only the low-performing ones *learned* better. Karen disagreed with the survey item "My students learn better working with a partner" (see Table 2) and wrote a note on her survey explaining that her higher ability students did not necessarily *learn* better. Prior to the program, all of the teachers indicated that they used cooperative learning strategies often and that their students were accustomed to working in groups. However, it became apparent during

TABLE 2. Student Survey Items and Attitude Mean Scores and Standard Deviations by Teacher

		Teacher					
Item		Ben	Barb	Karen	Kim	Total	SD
1.	I would like to learn more about geometry	2.31	2.15	1.88	2.07	2.02	.84
2.	Overall, I liked the program	2.00	1.98	1.81	1.98	1.95	.77
3.	I tried hard to do well in the program	1.35	1.46	1.33	1.36	1.37	.52
4.	I had personal control in deciding what to do in the program	2.39	2.05	2.00	1.97	2.11	.70
5.	I like activities where I have a chance to figure things out for myself	2.06	1.83	1.92	1.93	1.87	.78
6.	I learn better working with a partner	1.75	1.83	2.06	1.80	1.86	.97
7.	The program made me think a lot	1.53	1.62	1.44	1.52	1.53	.68
8.	I wish I had more help from the instructor during the program	2.38	2.52	2.76	2.38	2.48	.90
9.	What did you like best about the program?						
10.	What did you like least about the program?						
11.	How was this program different from your regular classroom?						

the program that students rarely worked in groups. Perhaps the teachers assigned students to work in groups, without emphasizing widely held cooperative learning principles (Johnson & Johnson, 1994, p. 79), thus making the quality of the interactions seem different from group work to the students.

Ben, although seemingly regimented and organized in his personal life, had a relatively loose class management style. His classes were fairly unruly but respectful. Ben demonstrated an affection and respect for his students. A self-described traditionalist, who is trying to change, Ben readily admits that he teaches the way he was taught. The problem as he sees it today is that students are less motivated and disciplined than a generation ago and the standards-driven curriculum is less relevant to students today. Therefore, he views the Sketchpad activities as "a more palatable way to approach some of the geometry concepts."

Ben was very active during the entire program. He continually tried to help students and would check students' activities booklets. He tried to anticipate problems and was very patient. To his credit, he seemed to fight his urging to be a bit too "helpful." He would try to ensure that students got the correct solution to a problem rather than guide students to proceed in small increments, allowing them to test erroneous hypotheses (as suggested by Land & Hannafin, 1996), which, of course, was understandable, even predictable, given lack of training. When confused, students often stalled in the program and waited for him to help them. It is interesting to note that Ben strongly disagreed with item 3 on the teacher survey, "I felt uneasy about letting my students control their own learning," while his students reported they had relatively less control (relative to other student participants) during the program. This probably reflects Ben's expressed willingness to embrace a more student-centered approach; but, in practice, he probably felt somewhat uneasy letting his students control their own instruction. Ben learned the software reasonably quickly and indicated he would try it again on his own.

Kim and Barb would have difficulty, in our view, breaking from their traditional roles. Kim treated her students much the way one might expect to treat lower elementary students. For example, several times during the study, she would chastise the entire class: "There is too much noise in here; everyone face the front, feet in front of you, and place your hands on your laps where I can see them! Now that's better; resume working quietly." Her students did not display animosity toward her. They simply seemed bored and disinterested. At times, Kim's feedback was almost confrontational, responding to student requests for help: "You know *that*. Look at the screen. What do you see?" and "Oh, no, the program isn't wrong. *You* are wrong!" One comment that perhaps reveals a great deal about her view about teaching was: "They've got to be able to use me as a facilitator, and they've got to be determined that they're going to change their outlook."

Kim was positive about the program's effectiveness and observed positive interactions among her students: "I saw the kids using a guess-revise-check method . . . they would check by trying it themselves on the computer." Kim had difficulty learning how to use Sketchpad and was uncomfortable answering any student questions that were software-specific. She reported after the program that she did feel more comfortable with Sketchpad, but not enough to use it again completely on her own. Before the program, she was a bit concerned about discipline problems with allowing students to work in pairs but was pleasantly surprised with the outcome.

Barb's classes were a bit unruly. Her concern for the students was not always evident. She talked about adopting new strategies and of integrating technology into her teaching but did little to learn the Sketchpad when she had the opportunity during the sessions. She graded papers or did other paperwork during the program, occasionally raising her head to discipline a student. She was embarrassed that her students didn't know the content and confessed that, on one occasion, she reviewed some of the program's instructional activities with her students before class despite agreeing not to do so. Barb reported that she liked the program and thought her students "got a lot out of it." She believed that pairing learners heterogeneously by ability was very effective.

Program Features. Ben responded to the open-ended survey question "What did you like *best* about the program?" that the program's measuring capability was his favorite part; Barb said the researchers' knowledge of the program was what she liked best; Karen liked the instructional booklet containing the 16 activities; and Kim liked the fact that students could discover things for themselves. Three of the four teachers' responses to the survey question "What did you like *least* about the program?" focused around technical difficulties with the program: Ben disliked a particular feature of the program (displayed geometric points had dimension which led to student misconceptions); Karen thought the program got tedious when students had to continually re-measure angles with each activity; and Kim expressed concern about the technical difficulty some of her students experienced in using the Sketchpad. Barb disliked the lack of timely feedback (which was, of course, intentionally withheld).

Teachers consistently expressed that students working on their own and managing their instruction was the main difference between the Sketchpad experience and their regular classroom. It is interesting to note that 24 students reported that working in teams was different from their regular classroom, which contradicted what the teachers reported to the researchers in advance of the program. The teachers all indicated that they used cooperative learning groups on a fairly regular basis.

All four teachers indicated they would use Sketchpad again and that they would use it in much the same manner as it was used in the present study. All but Ben reported that, while they felt more comfortable using Sketchpad, they needed practice using the tool. All teachers reported they thought their students developed a deeper understanding of the content using the Sketchpad activities than they would have using regular instructional practices. Responses to the final survey question "How often do your lesson plans include computer use?" ranged from once per month (Kim) to once or twice per year (Ben). Karen indicated that she uses computers very infrequently because she "cannot figure out how to manage it."

The Second Research Question

> 2. *What are students' perceptions and reactions to learning geometry in less structured environments with dynamic geometry programs like the Sketchpad?*

The student survey mean scores are summarized in Table 2. Student attitudes as measured by the post-program survey were generally positive. The most positive responses were to item 3, "I tried hard to do well in the program" ($M = 1.37$), and to item 7, "The program made me think a lot" ($M = 1.53$). This is consistent with the individual teacher responses to the parallel item on their post-program survey (Table 1), with all four teachers either agreeing or strongly agreeing that their students tried hard during the program and that the program made their students think a lot. Easily the most negative student response ($M = 2.48$) was to survey item 8, "I wish I had more help from the instructor during the program." This probably reflects the novelty of students working on their own and being able to figure things out for themselves. But it also points out that students are quite willing to take on more responsibility for their own learning than many teachers probably realize.

Student comfort with working in the program with less supervision was also evident in some interesting correlations found between survey items. Items 2 and 7 were correlated ($r = 150, p = .024$), indicating that the more the program made students think, the more they liked the program. Items 2 and 8 were negatively correlated ($r = -.243, p < .001$), where the more students liked the program, the less help they wanted from the instructor. Finally, item 5 was negatively correlated with item 8 ($r = -.138, p < .038$), indicating that the more students liked figuring things out for themselves, the less they liked help from the instructor.

It is interesting to note the pattern of differences in survey responses between Karen's and Ben's students. Karen's students reported they wanted less help during the program, had more personal control of their learning during the program, and wanted to learn more about geometry than did Ben's students. These differences in attitudes are consistent with, and may be partially explained by, the different teaching styles. In his regular classroom,

Ben's tendency to provide instant and direct feedback was limited in the present study, forcing his students to work independently. This style may contribute to a behavior of learned helplessness among his students. Throughout the semester, Ben's students may have become dependent on his help and experienced some level of discomfort when it was withdrawn.

Student responses to the open-ended question of what they liked *best* and *least* about the program are summarized in Table 3; student responses to the question "How was this program different from your regular classroom?" are reported in Table 4. Most students (120) liked drawing and creating the on-screen sketches the best (Table 3), which is consistent with the feeling of empowerment reported earlier. Fifty-one students reported that they disliked "nothing" about the program, and 44 disliked the booklet and reading the instructions. Clearly, the novelty effect of working on the computers overpowered other student responses (Table 4), with 84 students citing "working

TABLE 3. Frequencies of Student Likes and Dislikes About the Program

Student Response	Liked Best	Liked Least
Drawing/creating/constructing	120	
Working on the computer	27	
Working with a partner	14	
Learning new things	11	
Everything	9	
It was fun	9	
Hands on/different	8	
Nothing	7	
Other	21	
Nothing		51
Booklet/instructions		44
Writing		24
Identified specific activity/area		24
Partner/working with partner		14
Too difficult		11
Figuring things out		10
Other		34

TABLE 4. Frequencies of Student Responses to How the Program Differed from Their Regular Classroom

Student Response	F
Work on computers	84
Learned more/more interesting	29
Work with partner	24
It was fun	23
Being able to control own learning	17
Other	21

on computers" as the biggest difference between the Sketchpad program versus working in their normal classroom routine.

CONCLUSIONS, LESSON LEARNED, AND LIMITATIONS

Overall, both the teachers and students reported they liked working with Sketchpad. Teachers were quite positive about both their own experiences and their impressions of how their students fared during the program. In particular, they were encouraged by the level of engagement demonstrated by their lower-ability students and saw real potential in using more student-centered activities in the future. Students liked the hands-on nature of the activities and working independently of their teachers.

Although all four teachers reported they would likely use Sketchpad again "in much the same manner as we used it here," we are less optimistic. There are a number of barriers they would need to overcome. In our view, Kim and Barb, for example, lacked the technical skills and/or confidence in their skills to manage the instructional program. And if they did use it, it would likely not be to facilitate student-centered inquiry. The discouraging fact is that they did not seem willing to change their teaching style much at all, even though they expressed excitement about the potential outcomes in such environments. Consistent with Tobin and Dawson's (1992) argument, the pressure to cover content and improve standardized test scores, along with the many other powerful cultural influences brought to bear, will likely outweigh the short-term enthusiasm from the present experience.

The prospects for Kim and Ben, however, are a bit more encouraging. Both took an active interest in learning the tools in Sketchpad and seemed willing and able to adapt their style. It is likely that each will use Sketchpad with similar activities again. But how much their beliefs actually *changed* is doubtful. We were able to measure, with reasonable confidence, the teachers' beliefs about student-centered learning environments after the study, but we can only speculate as to how much/whether those beliefs have changed, since we had no pre-study data. Notwithstanding that limitation, we concluded that the attitudes of the four teachers we observed did not change much at all. Both Kim and Ben seemed predisposed to, and comfortable with, the notion of allowing their students to control their learning. The better question now seems to be not *how*, but *if*, beliefs can be changed. Saye (1998) reported that high school teachers who were accepting of ambiguity and unresolvable situations in their personal lives were more apt to be effective facilitators of technology-mediated, student-centered learning environments. Hannafin (1999) also found that experienced teachers' attitudes about learning and the classroom learning environments were very difficult to change. More research is needed to identify if, and under what circumstances, beliefs about learning can be altered.

Student reactions to learning in the less structured environment were quite positive. Students were comfortable with their new-found independence and, in general, with not having a prescribed instructional course charted for them. It is interesting to juxtapose Kim's and Barb's reticence with their students' willingness to accept ambiguity and uncertainty. Frustration and boredom did take a toll; but for the most part, students, even low-ability students, worked very hard and stayed on task. This is a potentially important finding, as it supports the notion that low-ability students, when provided with an ample amount of what Hannafin, Land, and Oliver (1999) call "metacognitive scaffolding," can succeed in open environments and need the structure of highly directed learning experiences. Surely, these students needed the structure provided in the activities, but at the same time, they were pushed and challenged enough to sustain interest.

We learned several valuable lessons that would be helpful for teachers and designers of similar instructional programs. First, the structure provided by the activities was critical in helping students focus and identify important areas. Second, teachers did not find Sketchpad as easy to use as we had anticipated, which posed challenges. Supplementary formal training would certainly be desirable, especially to master some of the more advanced features of the program. Although beyond the scope of this investigation, training teachers in different pedagogical strategies to facilitate student-centered inquiry may be beneficial. However, for reasons argued earlier, we are not convinced that training will be very effective.

Karen's desire to present some content in advance of the Sketchpad sessions is fairly representative of the basics-first approach described by the CTGV (1992). That Karen, who was the most successful at facilitating, valued this approach surprised us. There is very little evidence that either confirms or refutes whether basics first is superior to other, more open, student-generated strategies. Certainly further investigation is warranted in this area.

REFERENCES

Bielawski, L., & Lewand, R. (1991). *Intelligent systems design: Integrating expert systems, hypermedia, and database technologies.* New York: Wiley.

Choi, J., & Hannafin, M. J. (1995). Situated cognition and learning environments: Roles, structures, and implications for design. *Educational Technology Research & Development, 43*(2), 53-69.

Cognition and Technology Group at Vanderbilt. (1992). The Jasper experiment: An exploration of issues in learning and instructional design. *Educational Technology Research and Development, 40*(1), 65-80.

Cognition and Technology Group at Vanderbilt (1993). Anchored instruction and situated cognition revisited. *Educational Technology, 33*(3), 52-60.

Dunham, P. H. (1993). Does using calculators work? The jury is almost in. *UME Trends, 5*(2), 8-9.

Elsom-Cook, M. (1990). Guided discovery tutoring. In M. Elsom-Cook (Ed.), *Guided discovery tutoring: A framework for ICAI research* (pp. 3-23). London: Paul Chapman Publishing.

Goldenberg, E. P., & Cuoco, A. (1996). What is dynamic geometry? In R. Lehrer & D. Chazan (Eds.), *Designing learning environments for developing understanding of geometry and space* (pp. 351-367). Hillsdale, NJ: Erlbaum.

Hannafin, M. J. (1992). Emerging technologies, ISD, and learning environments. *Educational Technology, Research and Development, 40*(1), 49-63.

Hannafin, M. J. (1996). *Technology and design of open-ended learning environments.* ITFORUM [electronic listserv]. Athens, GA: The University of Georgia. Available: *http://tech1.coe.uga.edu/ITFORUM/home.html*

Hannafin M. J., Land, S., & Oliver, K. (1999). Open learning environments: Foundations, methods, and models. In C. Reigeluth (Ed.), *Instructional-design theories and models: A new paradigm of instructional theory* (pp. 115-140). Mahwah, NJ: Erlbaum.

Hannafin, R. D., & Freeman, D. J. (1995). An exploratory study of teachers' views of knowledge acquisition. *Educational Technology, 35*(1), 49-56.

Hannafin, R. D., & Scott, B. N. (1998). Identifying critical learner traits in a dynamic computer-based geometry program. *Journal of Educational Research, 92*(1), 3-12.

Hannafin, R. D. (1999). Can teacher attitudes about learning be changed? *The Journal of Computing in Teacher Education, 15*(2), 6-12.

Hooper, S. (1992a). Effects of peer interaction during computer-based mathematics instruction. *Journal of Educational Research, 85*(3), 180-189.

Hooper, S. (1992b). Cooperative learning and computer-based instruction. *Educational Technology Research and Development, 40*(3), 21-38.

Johnson, D. W., & Johnson, R. T. (1994). *Learning together and alone* (4th ed.). Boston, MA: Allyn and Bacon.

Massachusetts: A Paramount Communications Company, Key Curriculum Press (1993). *The Geometer's Sketchpad*. [Computer program]. Berkeley, CA.

Lakoff, G. (1987). *Women, fire, and dangerous things: What categories reveal about the mind*. Chicago: University of Chicago Press.

Land, S., & Hannafin, M. J. (1996). A conceptual framework for the development of theories in action with open learning environments. *Educational Technology Research and Development, 44*(3), 36-53.

Meyer, D. K. (1992). What is scaffolded instruction? Definitions, distinguishing features, and misnomers. *National Reading Conference Yearbook* (No. 42, pp. 41-53).

Means, B., (Ed.). (1994). Introduction: Using technology to advance educational goals. *Technology and education reform: The reality behind the promise* (pp. 1-21). San Francisco: Jossey-Bass.

Palincsar, A. S., & Brown, A. L. (1984). Reciprocal teaching of comprehension-fostering and comprehension-monitoring activities. *Cognition and Instruction, 1*, 117-175.

Patton, M. Q. (1990). *Qualitative evaluation and research methods*. Newbury Park, CA: Sage.

Pea, R. D. (1993). Practices of distributed intelligence and designs for education. In G. Salomon (Ed.), *Distributed intelligence* (pp. 88-109). New York: Cambridge.

Rieber, L. P. (1992). Computer-based microworlds: A bridge between constructivism and direct instruction. *Educational Technology Research and Development, 40*(1), 93-106.

Rogoff, B., & Gardner, W. P. (1984). Adult guidance of cognitive development. In B. Rogoff & J. Lave (Eds.), *Everyday cognition: Its development in social context* (pp. 95-116). Cambridge, MA: Harvard University Press.

Saye, J. (1997). Technology and educational empowerment: Students' perspectives. *Educational Technology Research and Development, 45*(2), 5-24.

Saye, J. (1998). Technology in the classroom: The role of dispositions in teacher gatekeeping. *Journal of Curriculum and Supervision, 13*(3), 210-234.

Schank, R. C., & Cleary, C. (1995). *Engines for education*. Hillsdale, NJ: Lawrence Erlbaum.

Scott, B. N. (1998). *Curricular change in higher education: What we say and what we do*. (Eric Document Reproduction Service No. 415 779)

Singhanayok, C., & Hooper, S. (1998). The effects of cooperative learning and learner control on students' achievement, option selections, and attitudes. *Educational Technology Research and Development, 46*(2), 17-32.

Tobin, K., & Dawson, G. (1992). Constraints to curriculum reform: Teachers and the myths of schooling. *Educational Technology Research and Development, 40*(1), 81-92.

Vygotsky, L. S. (1978). *Mind in society*. Cambridge, MA: Harvard University Press.

Young, M. F. (1993). Instructional design for situated learning. *Educational Technology Research and Development, 41*(1), 43-58.

Michael L. Connell

Actions on Objects:
A Metaphor for Technology-Enhanced
Mathematics Instruction

SUMMARY. When thinking of reasoning, problem solving, communication, and connecting related ideas, the tool of choice in nearly every discipline is the microcomputer. Furthermore, unlike the traditional calculator, the modern classroom computer has an unparalleled ability to implement both graphical and procedural components of mathematics understanding in a single unified object. By students' creation and utilization of mathematically relevant computer-based objects, this dual encapsulation provides them with a unique opportunity to see both the form of representation and their actions utilizing this representation simultaneously.

This paper suggests that the object-oriented environments that modern technology enables are ideally suited to parallel and facilitate the ability of students to take a broader variety of *action upon objects* of a nature and kind hitherto unknown. These student-controlled *actions* upon these mathematically powerful and computer-enabled *objects* have the potential for creating classroom environments that both surpass the pale hopes of the integrated learning system and surprise those wedded to a conservative view of Piagetian developmental levels. *[Article copies available for a fee from The Haworth Document Delivery Service: 1-800-342-9678. E-mail address: <getinfo@haworthpressinc.com> Website: <http://www. HaworthPress.com> © 2001 by The Haworth Press, Inc. All rights reserved.]*

MICHAEL L. CONNELL is Associate Professor, College of Education, University of Houston, 344 FAH, Houston, TX 77204-5872 (E-mail: Mkahnl@aol.com).

[Haworth co-indexing entry note]: "Actions on Objects: A Metaphor for Technology-Enhanced Mathematics Instruction." Connell, Michael L. Co-published simultaneously in *Computers in the Schools* (The Haworth Press, Inc.) Vol. 17, No. 1/2, 2001, pp. 143-171; and: *Using Information Technology in Mathematics Education* (ed: D. James Tooke and Norma Henderson) The Haworth Press, Inc., 2001, pp. 143-171. Single or multiple copies of this article are available for a fee from The Haworth Document Delivery Service [1-800-342-9678, 9:00 a.m. - 5:00 p.m. (EST). E-mail address: getinfo@haworthpressinc.com].

143

KEYWORDS. Activity theory, constructivism, mathematics, student cognition, object reification, learning theory

The potential inherent in the modern microcomputer to serve as a bridge between and among instructional theories has not been lost upon many researchers in mathematics education and curriculum theory (Brownell & Brownell, 1998; Garofalo, Shockey, & Drier, 1998). The computer easily enables a dynamic and active learning environment within which each of the process strands of the National Council of Teachers of Mathematics (NCTM) standards might find a home (NCTM, 1989). After all, when one thinks of reasoning, problem solving, communication, and connecting related ideas, the tool of choice in nearly every discipline is the microcomputer.

Furthermore, unlike the traditional calculator, the modern classroom computer has an unparalleled ability to implement both graphical and procedural components of mathematics understanding in a single unified object. By students' creation and utilization of mathematically relevant computer-based objects, this dual encapsulation (Sfard, 1991) provides them with a unique opportunity to see both the form of representation and their actions utilizing this representation simultaneously. For this reason alone, the computer would be a natural tool for both classroom use and theoretical musings.

This has not been lost upon teachers, and, despite those who would use the computer solely for the presentation of pre-packaged instructional units, there is a growing consensus that the computer can be an ideal tool for knowledge construction at an individual or group level (Harvey & Charnitski, 1998). As we explore the potential for classroom uses of computer technologies, we have a once-in-a-lifetime opportunity to blend the best of learning theory and the practical realities of student actions.

We will illustrate how the object metaphor, found in modern window-based operating systems and programs, transfers to *action upon objects* models of mathematics teaching and learning. In particular, we suggest that the object-oriented environments that modern technology has created are ideally suited to parallel and facilitate the ability of students to take a broader variety of *action upon objects* of a nature and kind hitherto unknown. These student-controlled *actions* upon these mathematically powerful and computer-enabled *objects* have the potential for creating classroom environments that surpass the pale hopes of the integrated learning system and would surprise those wedded to a conservative view of Piagetian developmental levels.

THE REAL WORLD OF THE VIRTUAL

Let us examine a rather different role the computer might play as it relates to objects of knowledge: the notion of a student creating a computer object which then becomes an object upon which further instantiated actions could then be taken. This fits in well with both emergent and generative theories (Sfard, 1994; Fagan & Thompson, 1989) and, once a few initial conceptual hurdles are cleared, is both amazingly simple to understand and powerful in its implementation.

Given the long history of cognitive science for borrowing the most current metaphors from computer science, it should come as no small surprise to see the applicability of many of the object-oriented programming strategies to reflect aspects of human cognition (DuPlessis, 1995). And, indeed this new metaphor plays out very well for the case of mathematics education. The tools supported by the modern computer enable a new class of *objects with which to think*.

In order for this to happen, however, we must attempt to leave behind any preconceived notions about the role of the technology as being most useful for information presentation and delivery. A much healthier perspective, at least in terms of understanding this new approach, would be to envision the technology as being used to provide a *milieu* within which knowledge construction can occur (Connell & Abramovich, 1999). To see what this might entail, let me begin by telling a brief story from a very early research study using the Windows-based authoring program ToolBook.

I had created a ToolBook program that I called a "Cognitive Playspace" for children to explore various foundational notions of mathematics. My goal at the time was to examine the extent to which traditional physical manipulatives might be augmented by technology. One activity, in particular, involved the use of various geometric shapes. I had intended these to be analogous to the traditional pattern blocks with which the children were already familiar, but not to be identical to them. In this activity, the children were presented with a variety of puzzles to solve.

In the puzzle set for a particular day, one of the tasks was for the child to reconstruct a pattern made, using the provided shapes, by manipulating geometric forms on the computer screen. I had developed some fairly simple tools that would enable the geometric shapes to be rotated, translated, and generally moved about the screen (see Figure 1).

Of particular interest for the notion of object creation is an event that took place when I brought a group of kindergarten and first-grade students to the university to work with this exploration package.

Prior to this particular occasion, the children had other experiences in using both ToolBook and the relatively primitive Windows machines. By the time of this incident, the children typically had very little difficulty relating to the computer itself. In considering their level of expertise, I would say they were easily the equal of a computer-savvy kindergarten student of today. I should mention that we were using 386 and 486 computers to run Windows for workgroups, which, at the time of this study, represented an extremely high-

FIGURE 1. Initial student work using Cognitive Playspaces.

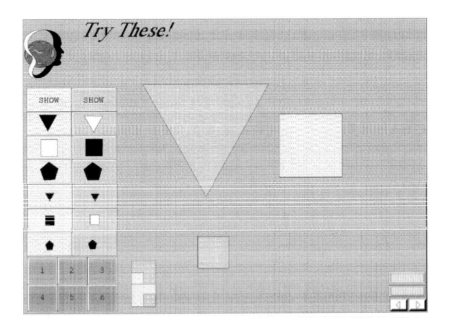

end product. Furthermore, for the type of software that we were running, these computers were more than adequate, even by today's standards.

As an aside, the notion of using the most powerful technology and authoring system then available to work with kindergarten students struck many as being amazing. I made the point then, which I still believe, that, in order for the computer (of that day) to be meaningful, it *required* the highest end product. One of the major pleasant surprises of the last nine years is how well these earlier findings hold up. Indeed, the off-the-shelf machines of today are quite comparable (in terms of the visualization and manipulability of objects they engender) with these early efforts. Not only was this work ahead of the curve, it was outside the box totally.

But what I remember most, relative to the notion of object reification, came from the experiences of one young man who, in his manipulation of the graphic objects, managed to drag all of them off of the computer screen. This was a bit more problematic than it might first appear, for in the nature of "it's not a bug, it's a feature," I had not created a simple button to return objects to the screen once they had been dragged off the working area.

I must confess that this was a complete oversight on my part. The program was under development, and I had not yet considered what would happen if objects were dragged off the screen. In the course of developing the activities, I had simply clicked to the next screen and then back to reset the screen. Of course, this child was unaware of that strategy. When he dragged them off of the screen, for all he knew they were gone forever.

I'll never forget him looking up at me, his eyes swelled with tears, as he said, "Help me, Dr. Connell! I have lost all of my toys" (see Figure 2).

His painfully sincere statement serves as a powerful reminder that, in the mind of this child, these computer images represented *real* objects upon which he was carrying out *real* operations. Furthermore, these operations carried with them a heavy kinesthetic component that was totally unexpected at the time. Subsequent observations and interviews illustrated that the physical manipulation of his eyes and hands with the motions of the mouse, as translated to the computers screen, were in his mind a series of actual actions upon real objects.

FIGURE 2. Screen without "toys."

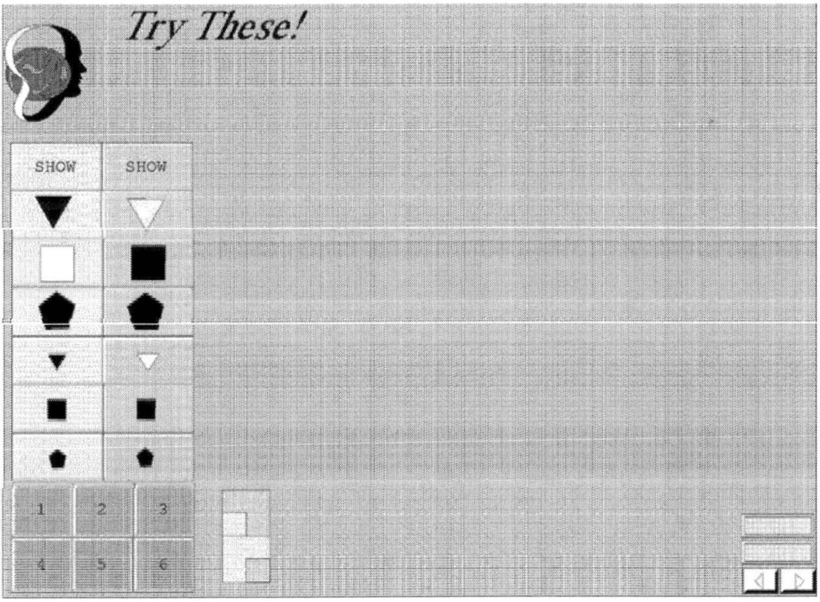

To put it in more poignant terms, this was no virtual reality to this child. These were his real toys, and they were lost. I'll never forget his joy as I showed him how to get his toys back.

The experiences of this child that the technology had engendered were extremely powerful for him. Furthermore, they proved to be similar to the real-world pattern blocks from which the geometric shapes had been designed. The combination of kinesthetic motion required in manipulating the objects with the mouse and the geometric shapes on the computer screen appeared to provide a direct analog to the kinesthetic motion required to move the pattern blocks on the desktop. To these children differences between moving the virtual object on the computer screen and moving the physical object of the pattern blocks on the desktop were so small as to be indistinguishable.

For this child, this represented a valid action upon a real object. To the degree that mathematics represents a series of actions upon reified objects of growing complexity and abstraction, this then provides an excellent clue in terms of a starting place for effective technology use.

THOUGHTS ON SYMBOLIC COMPUTERS
AND MATHEMATICAL OBJECTS

Despite much current usage, technology in the mathematics classroom is best suited not for presentation of abstracted symbols, but rather as a tool for exploring the foundational objects upon which the mathematics is to be built and, in some cases, by providing mechanisms for varying the underlying nature of the object upon which the symbol is constructed so as to elicit the formation of selected concepts. This is a profoundly significant point. Typically what we see in mathematics education is for the technology to be used solely for its symbol-processing power–not as a tool to develop and explore the foundational attributes of mathematical objects. Yet, in looking at the psychological construction of mathematical meaning, we observe that meaning is best made by students' performance with actual actions upon actual objects. In order for this to occur, the object must become real in the minds of the students. It must become instantiated. It must possess well-defined attributes in the mind of the child so that the eventual symbol generated by the experience has well-defined properties.

A particular item of interest has to do with the nature of the action and the nature of the object in the ongoing construction of mathematical concepts. Whitehead (1978) in "Process and Reality" argues quite convincingly that it is through the increasing sophistication and power of abstraction, enabled through symbol systems based upon more primitive objects–such as the algebra–prevalent in mathematics, that we are able to leverage forward our thinking. There is thus a long history regarding the nature of the foundational object upon which one acts and the importance of this foundational object in the generation of later conceptualizations.

Piaget, of course, takes this much further with his own firmly developed notions regarding the stages through which these dynamic constructions are occurring. I find, however, that Piaget is somewhat limiting in terms of his application to mathematics. A strict reading of Piaget, for example, would lead one to the conclusion that young children are incapable of abstract mathematical thinking. However, when they are given the correct models and tools to think with, students are capable of developing sophisticated representations and understandings of the mathematical concepts at a level beyond which Piaget would recognize.

I have come to believe that good thinking is good thinking regardless of the level it happens to take. As a part of the approach to learning and instruction that is described here, I have come to realize that, whether dealing with pre-kindergarten to third, fourth through sixth, middle, high, or graduate students, we are really looking at the generation of concepts through a series of activities based upon developmentally appropriate objects of thought. For the case of the elementary and the pre-kindergarten students, these objects of thought are expressed in terms of real world-objects or manipulatives. As such, these objects have the potential of carrying a great deal of information, some that is mathematical, some that is not.

There are some significant differences between a traditional manipulative and a computer-based manipulative. A good traditional manipulative also has properties that simultaneously present to the students opportunities for foundational concept creation. This is a hallmark of most of the classical manipulatives presently adopted within the field. A traditional manipulative–such as the Dienes blocks–also illustrates concepts other than those for which they were originally created. A set of $Base_{10}$ blocks, for example, can simultaneously present to the student experiences with mass, density, smell (senses), etc. In other words, we have the potential to generate more than just the mathematical topic of the day. In some ways, this is a very good situation to have, as it provides much flexibility and great linking power in a given manipulative.

Of course, this multifaceted presentation of experience to the senses is not necessarily a good thing. It is possible that these additional added features are actually seductive details that distract from the core concepts that we would hope are being developed. This leads to the observation that the narrowing of focus brought about via a sketch is the first step toward the later highly abstracted representations such as algebra. Whitehead (1978) in "Process and Reality" states "the simplicity of clear consciousness is no measure of the complexity of complete experience" (p. 267).

A computer-based manipulative best serves as a visual representation of a predetermined concept. In particular, one of the key differences between a computer manipulative and the real-world manipulative lies in the degree of abstraction that occurs due to the use of the computer to generate the object of thought. In the case of mathematics, quite often we are trying to teach a specific representation as opposed to broad multiple uses of the manipulative. Thus as a domain we utilize standard representations such as Dienes blocks, which are specifically constructed to carry a single meaning at the expense of other potential meanings that the material might be used for.

A CASE FOR STRONG CONCEPTUALIZATION

Due in part to this increased abstraction inherent in many computer-enabled objects, it is extremely important for children to develop strong concepts. There should be well-developed understandings regarding what the objects they are manipulating are to be used for mathematically and how these objects are appropriately and inappropriately used. One analogy that comes immediately to mind is the *smart wizards* seen in many Microsoft applications. These wizards are often helpful, but without a strong understanding of the underlying rules governing their behavior, their suggestions can be confusing. Projecting this type of object into mathematics, it is easy to envision a case where these wizard-enhanced objects could have profound effects upon mathematics learning.

To see how this might play out, let's imagine that there are two quantifiable sets that we wish our students to explore. Let's further imagine we have identified various operations that are possible to use on these two sets and have created a series of technologically enhanced objects to use in their exploration. In this scenario, it is possible that a beginning student might select a *set addition* object, with its associated wizard, using the operating software, and give it the instruction to combine these two sets. The *set addition* object–which has more than a bit of *intelli-sense* and *wizardry* (to use Microsoft terminology for a moment)–does the cybernetic equivalent of looking at the task ahead of it and then responding to the student, "Are you sure that you really want to do this?"

In order to be successful in this new environment, the child has to know whether or not this really is an appropriate operation to perform upon these objects, whether or not he or she has asked the right object to do the job, and how to interpret the eventual results.

This is not a fanciful example. Such scenarios are becoming all too common and indeed become more than a bit of a nuisance by frequently showing up in *intelli-sense* technologies. For example, you can be writing a letter and, before you can even finish the first paragraph, the Wizard de jour will pop-up. "Hi there! It looks like you're trying to write a letter. How about some help?" Typically I really don't want or need the help because it doesn't fit with either my writing style or the way I want to put this on paper. After all, an academician writes differently all the time. For whatever reason, however, it's important to note that, in this scenario, I am the expert. I can override the suggestions of any object or wizard when it's not in line with the tasks that I need to have done.

In this new world I'm envisioning, however, of computer-enhanced mathematics and mathematics instruction via active objects, this may not always be the case. Let's imagine that we are doing some integration and a wizard makes a suggestion on the boundary conditions, which nine times out of ten would be right. If your problem happens to fit the tenth condition and you succumb to the wizard's advice in the face of your own lack of concept, you are in major trouble.

This problem occurs on a daily basis in computer programs commonly used in the statistical analyses of data. A very real problem has occurred as more and more researchers are getting access to higher and higher levels of statistical programs. In many cases programs such as SPSS and SAS will enable processing beyond the interpretive levels of the users. It is very common to find data sets that are spherical and never checked, and post-op comparisons that are performed correctly but selected inappropriately–all of this because the tools given for the individual to think with were, in many ways, smarter than the people thinking with them. This trend is one that shows every tendency of continuing and accelerating.

This plays out with a vengeance in the educational arena. We are already able to design and implement intelligent objects with more "number-sense" than the beginning students who will be utilizing them. We are very close to being able to come up with objects to think with that are more intelligent than the people working with them. This is not intended in a callous or mean fashion. We don't expect a tremendously high level of mathematical metacognitive knowledge at the first-grade level. After all, the learner is just putting all this stuff together–in many cases for the very first time. So, to return to our earlier example, it would be very easy to imagine a *set addition* object that basically says, "Take any two numbers you give me and combine them using the operation of addition." In terms of sheer processing power, this could easily be at a higher skill level than the child using the object.

If we are to be effective teachers in this new technology-enhanced environment, we need to assure that our students truly understand the concepts. If this is not done, all of our lovely *correct* answers are meaningless. This was a major concern in the calculator-based reform effort of 15 years ago; it is even more crucial in the computer environment. Let us see why this should be the case. Using a traditional calculator, you still had to plug everything in yourself–much like the old command-line DOS interfaces or line-by-line BASIC compilers. The new computer environments and many of the newer calculators are becoming increasingly object-oriented. Therefore, it is entirely possible that we may end up having the terminal smarter than the user. I have always thought we were better off having dumb terminals and smart users in computing. Without extreme care, we will soon be facing the reverse.

EXAMINING THE *TOOLS TO THINK WITH*

The creation of new *objects of thought* or *tools to think with* can become very powerful pedagogically, assuming we understand the concepts underlying them. The hidden danger surfaces when we cannot understand the underlying mathematical concepts upon which the active objects are operating, and simply take them for granted, follow their recommendations blindly, and accept their results at face value.

This would be analogous to letting your writing be totally edited by wizards in your word processor. The following poem, which has enjoyed wide popularity among information technology faculty through the years, will serve to illustrate the dangers of such an approach.

SPELL BOUND

I have a spelling checker,
It came with my PC.
It plainly marks four my revue
Mistakes I cannot sea.
I've run this poem threw it
I'm sure your pleased too no,
Its letter perfect in it's weigh,
My checker tolled me sew.

Because of our in-depth knowledge of words and word usage, it is easy to see the humor in this piece of writing. The errors are obvious and mostly harmless. If, however, we are looking at a computational object that was created with corresponding flaws in underlying logic, the errors would not be nearly so obvious, harmless, or humorous. Consider one such object, shown in Table 1.

In this example, it may not even be possible to identify the purpose for which the "object" was created, the assumptions that underlie its creation, or the use to which its results might be applied. The calculations are done

TABLE 1. A Mathematical Object Without Context

Depth	The Total Relation	Diff. from 1	Radical Portion	Diff. from .5
1	2.00000000000	− 1.00000000000	1.00000000000	− 0.50000000000
2	0.00000000000	1.00000000000	0.00000000000	0.50000000000
3	1.08239220029	− 0.08239220029	0.54119610015	− 0.04119610015
4	0.99401088611	0.00598911389	0.49700544305	0.00299455695
5	1.00019859327	− 0.00019859327	0.50009929664	− 0.00009929664
6	0.99999679901	0.00000320099	0.49999839950	0.00000160050
7	1.00000002540	− 0.00000002540	0.50000001270	− 0.00000001270
8	0.99999999990	0.00000000010	0.49999999995	0.00000000005

correctly, of that we have no doubt; but to what use are they to be put? I am reminded of Douglas Adams' science fiction classic *Life, the Universe, and Everything* (Adams, 1983). In this book, we learn that the answer to all of the truly deep problems of philosophy, metaphysics, etc., is actually 42. The difficulty is that we do not know precisely what these questions are or in what form this answer fits.

A MODEL FOR TECHNOLOGICALLY ENHANCING MATHEMATICS LEARNING

A major theme that is emerging in this writing is that of *action upon objects*. This, in turn, leads to some foundational questions surrounding the nature of the objects and the types of actions that one might be expected to perform upon them. These *actions-upon-objects* models, for lack of a better term, are very powerful in both the laboratory as well as in the predictive power they enable in the minds of the students. I feel that they also capture quite a bit of current interest in the field, as evidenced by recent thinking on object reification (Sfard, 1994).

What I am trying to add to the mix is the notion of a firmly developed and articulated way of looking at what these objects might be and, in particular, how we might utilize them to develop mathematical thinking. I have developed an approach that results in students developing mathematical thinking, regardless of the developmental level and nature of the object. When this method is followed, we repeatedly observe markedly similar patterns of thought on the part of the students. This is all the more significant when we consider that this parallelism shows itself in the same type of thinking taking place at each developmental level.

Let me reconstruct some of my notions of an *action-upon-objects* model I have been using in my university classes and personal thinking. As you will note, my approach toward addressing these questions has been very heavily influenced by readings of and work with Mikhail Bounieav and Sergei Abramovich. Mikhail Bounieav and I have been developing a way of thinking about step-by-step development of mental activities as enhanced by technology (Bounieav & Connell, 1999; Connell & Bounieav, 1997). With Sergei Abramovich, I have been looking at the nature of the new tools to think with that technology provides (Connell & Abramovich, 1999). The development of this theme includes developing concepts through mental picturing and notions I have been developing over the last 20 years regarding actions upon objects of various types (Connell, 1986; Connell, 1988).

Background Examples

We see a growing consensus that the generation of mathematical thinking is best facilitated through a series of activities of various sorts that are carried out upon objects selected for their possession of certain specified properties that can instantiate mathematical concepts. Let's see what we mean by this. At the elementary level, the objects the children are capable of *thinking with* or *acting upon* are influenced by both their developmental level and their prior experience. In particular, we find that young children are not able to think with formalized abstracted mathematical objects. This should not be a major surprise, as it has been part of our understandings of human growth and development for some time.

This limitation, however, at first glance would appear to limit the degree of mathematics which might be made. As Whitehead correctly notes, it is when one reaches the abstracted levels of mathematical formalism that he or she can really leverage forward his or her thinking. It is at that point that the tremendous growth in the intellectual potential of the individual might occur. This has led many to speculate and even to promote the notion that a young child is incapable of rigorous mathematical thinking and, in fact, the young childhood level is basically preparatory for the *real* mathematics that the student will develop later.

Based upon my experiences in the classroom and my own theoretical musings, I have taken a very contrary position to this. As I have stated in many different venues, young children are capable of very well-developed mathematical thinking if the *objects with which they think* and the *questions upon which they think* are of an appropriate level and type for their developmental abilities. I will be the first to acknowledge that this is a different type of mathematics *content* than we often see in more formal mathematics, but the thinking strategies are in direct parallel to those exhibited at higher levels.

Let us work through a few examples to see how this might play out. For a preschooler working with pattern blocks, we can ask questions concerning these blocks that require elaborated thinking. Suppose the child creates a pattern composed of a square followed by a triangle and then a parallelogram as a base as shown in Figure 3.

The child can easily and correctly predict what comes next from this base as it is continued. Indeed, for any given base, children will quickly learn how to extend that pattern and to create their own patterns from bases of their choice as shown in Figure 4.

Now some might suggest that this type of thinking is more replication than prediction. However, we commonly see the same type of thinking occur when we observe algebra students using a guess-and-check strategy to fill in values in a function like that shown in Table 2.

I really believe that we are not just laying a foundation for later mathematical thinking. We are indeed actually seeing mathematical thinking that is appropriate for the *objects* that the student is able to *think with*.

The Sigma Grid

However, if this approach is to be effective, the objects to think with must be developmentally appropriate for the student. For the past 15 years, I have been researching and extending one such model that has proven to be very helpful in identifying the level of objects to think with and some of their properties (Connell, 1998; Connell & Ravlin, 1988; Connell, 1986). Figure 5 serves to illustrate this approach (Connell, 1986; Peck & Connell, 1991).

Let me provide a quick general overview of this approach. The instructional goal is to enable gradual student construction of meaning through the use of *manipulatives* through *abstraction* via four transitional *object* types

FIGURE 3. Pattern base.

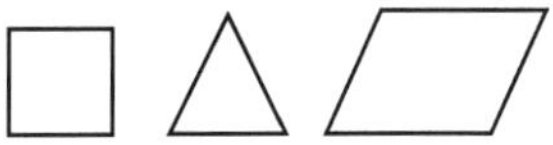

□ △ ▱ ╱

FIGURE 4. Extended pattern.

TABLE 2. A "Guess-and-Check" Function

X	F(x)
1	3
2	6
	9
4	
5	15
6	
7	21
8	
9	27
	30
11	33
.	.
.	.
.	.

FIGURE 5. Model of constructivism showing the *objects-of-thought* described.

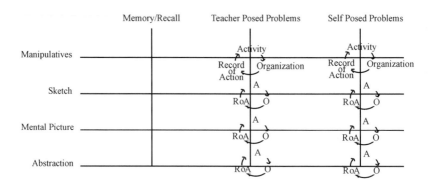

(Connell & Ravlin, 1988; Connell, 1988). Within each problem type, the student will typically encounter three types of activities:

1. *Memory/recall*–often of terminology;
2. *Teacher-posed problems*–related toward student construction of concepts;
3. *Student-posed problems*–based upon developing understanding of the problem space presented and its relations to other problem types.

These activities will be experienced by the student in the form of related problems requiring the use of the developmentally appropriate *object* to think with: *manipulatives, sketches, mental pictures,* and *abstractions.* Furthermore, at each location in the model where a student will encounter a problem–either teacher- or self-imposed–the student will solve this problem via activities that are then organized and recorded for later reference.

An extremely elementary example of this model showing actions upon all of the object types might include initial *actions* upon a *manipulative*–such as a pile of counters used to develop elementary addition. A *sketch* might then be drawn recording the actual counters, which in turn would serve as an *object of thought* for further construction of meaning. *Mental pictures,* in addition to serving as a further representation of the problem space, provide natural entry points for technology–which will be utilized in technology-aligned classrooms. *Abstraction* would occur when it is no longer necessary for the student to use countable counters but when that student is capable of reflecting upon the constructed representations in the construction of new knowledge, a process that Piaget referred to as "reflective abstraction." As we thus expand our earlier notions of action upon object we can see that we are working with a carefully selected set of developmentally appropriate primitive objects[1] and experiences with these objects to build up a working vocabulary and subsequent conceptualization.

Such a conscious development of vocabulary via actions upon objects fits in very well with Vygotskian notions as well as those found in the step-by-step development of mental activities approach (Connell & Bounieav, 1997). Furthermore, these basic actions upon objects leading to vocabulary development provide the student the opportunity to develop and use words in an externally referable context within which any questions or deficiencies might be easily observed by both teacher and student.

The development of a commonly shared vocabulary is an extremely important part of the instructional model shown in Figure 5. This is not a new idea, as the importance of a commonly shared vocabulary in teaching has been described for thousands of years. In Plato's retelling of the discussions between Meno's slave Anytus and Socrates, for example, one of the first questions asked by Socrates is: "Does the boy speak Greek?" To use the *speak Greek* Socrates metaphor, these experiences provide a context within which a skilled teacher or instructor can create meaningful problems.

In particular, I have argued that the Sigma Grid may be utilized with a parallel teaching strategy at each of the developmental levels–thus implying that the type of thinking which this engenders will likewise be parallel. In this model of teaching, we would expect to see evidence of well-developed mathematical thinking at each of the stages of *manipulative, sketch, mental picture,* and *abstraction.* Furthermore, as we extend our thinking of this approach, these developmental levels may also be thought of as describing the nature of the objects that students are able to utilize in their own thinking at their current stage of development. Our task is to ensure an appropriate match between the *objects with which one might think* and the *questions that we think about.*

Two Parallel Examples

Let us see how this model plays out for the case of a young child who is just beginning the process of acquiring basic number concepts. The initial task for the young child is to perform sufficient actions upon a foundational set of *manipulative objects* to develop a working vocabulary for later use. This vocabulary must include the terminology used for the manipulative itself, relevant properties of the manipulative, and the canonic problems for which the manipulative is used to explore.

Consider the Dienes $Base_{10}$ blocks as an example. Typically a child begins by using the blocks to build with–just as with any other set of building blocks. Through carefully guided activities, the young child will come to explore more of the mathematically relevant properties of the blocks and begin to assign the appropriate terminology to them.

Thus, the smallest block (see Figure 6) is soon recognized as a unit. Building from this basic foundation, the child learns that it is this unit that we *count* when using the $Base_{10}$ blocks. This is the primitive object serving as the source for later representations within this system of modeling.

From this beginning, other vocabulary relating to the blocks is carefully developed, such as Hundreds Flat, Tens Rod, and Thousands Cube (see Figure 7).

With this vocabulary in place, experiences are designed to explore the relationships between the numbers represented in these *objects to think with,* and problems are posed which require the child to consciously and strategically *act upon these objects* in order to solve them. The child will then pose problems of his or her own, which will end up involving further *actions* that the child will perform upon the primitive *objects* with which he or she has been working.

To see the parallelism of approach this method incorporates, let us contrast this example from early elementary mathematics with one from introductory

FIGURE 6. Unit.

trigonometry (see Figure 8). First of all, the objects these students are thinking with are quite a bit more abstract. Let us look at a technology-enhanced *sketch* object designed to allow exploration of the sine and cosine function via the unit circle. Just as was done with the young child, we begin by developing a working vocabulary looking at selected points along a unit circle.

FIGURE 7. Ten, hundred, and thousand.

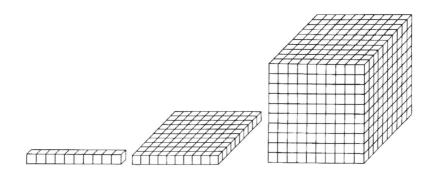

FIGURE 8. A unit circle trigonometry object.

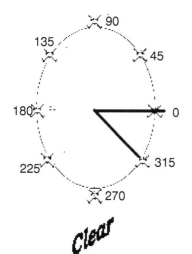

Theta	Sin Theta	Cos Theta
315	−0.7071	0.7071

Click Here for More

This *sketch* object is indeed more sophisticated and is best thought of as a dynamic sketch, due to the changing nature of its constituent parts. Despite this increased sophistication, however, we can observe the same pattern of thinking that the student uses in exploring his or her developing notions on trigonometric functions. The student must first acquire a working vocabulary regarding the object and some of its properties prior to any meaningful questioning or problem solving. Careful examination of the unit circle shows that there are two colored lines that are formed each time a ray crosses the unit circle at any particular angle, their lengths corresponding to the sine and cosine of the angle which is used. As the student explores various angle combinations, the associated sine and cosine are displayed in both numeric and graphic form, providing multiple representations for the same concept (see Figure 8).

Just as was done with the Base$_{10}$ blocks, once this vocabulary is in place, experiences are designed to explore the relationships between the functions represented in this technology-enabled *object to think with*, and problems are posed which require the student to consciously and strategically select *actions using this object* in order to solve them. The child will then pose problems of his or her own, which will end up involving further *actions* that the child will perform upon the *objects* in order to solve them. This same sequence likewise may be followed for the sketch and mental picture levels of the model.

What should occur next, regardless of the students' developmental level, would be for a skilled teacher or instructor to pose follow-up problems or questions relating to the newly instantiated and defined object of thought. This would again hold true whether we're talking about a physical *manipulative* object; a *sketch* object of predictive power, such as an interactive fractions object; or a *mental picture* object, such as a comparison of mass based upon remembrances of experience; or a formal and logically *abstracted* object; such as function or some other mathematical construct. In each case, we observe a skilled teacher using newly developed objects as a venue within which questions are to be asked and problem situations explored via student *actions* upon these very same *objects of thought*. As these examples serve to illustrate, these *objects of thought* become the basis upon which later mathematical thinking occurs.

GOOD THINKING IS GOOD THINKING, AT EVERY LEVEL AT WHICH IT OCCURS

It should also be noted that the nature and form of the thinking and reasoning strategies run parallel across each of these developmental levels. We can easily observe young children reasoning with the *manipulative* objects, communicating their findings with others using the *manipulative* objects, connecting their most recent experiences with previous experiences with the *manipulative* objects, and using these same *manipulative* objects in making quality judgments regarding their work. We can likewise easily observe the same strategies being used at each of the other developmental levels of objects within classrooms utilizing this approach.

If the question that has been posed relies upon *action* using *objects* that the child is capable of manipulating and has developed personally meaningful understandings for, then the child is typically successful in his or her problem-solving efforts. This holds true whether the form of the *action* upon the *object* is via a physical manipulation, a symbolic manipulation, or a more abstracted application of logical formalisms. It is important to note that I am not suggesting that we are necessarily observing extreme mathematical sophistication. However, what I am arguing for is that it is possible to observe a parallel form of mathematical thinking as students perform their respective *actions* upon their understood *objects* at each of these developmental levels. Thus, I am quite comfortable making the claim that we can observe quality mathematical thinking at the preschool level as well as at the graduate level.

Furthermore, as we extend this model, the children are given the opportunity to develop problems of their own based upon the objects that they recently defined, worked with, and for which they developed problem-solving skills and schemata. This is an important part of the instructional strategy, for without this piece of the puzzle, the children will always look to someone else to serve as the source of their problems and as final judge of the answers to the problems they face. This ability, to pose one's own problems and to then successfully solve these problems, provides further opportunity for growth in mathematical thinking and problem solving. In this, we also see once again the direct parallelism between each of these developmental stages.

I feel that this is an extremely important point. In a simple interpretation of Piaget, one might easily get the opinion that this parallelism would be impossible. After all, didn't Piaget state that the young child is incapable of abstract logical thinking? This does seem to preclude the possibility that the young child can utilize the same strategies as a more developmentally advanced student.

The path that I have chosen out of this seeming difficulty lies in the manner in which I view the mathematical actions of the students. I am not going to equate the ability to use an abstract formal logical structure with mathematical thinking. There are many students who can perform routine "plug and chug" operations utilizing algebra–despite little or no ability in mathematical thinking and problem solving. Conversely, it is very common for a young child to be capable of quite sophisticated chains of mathematically relevant reasoning–provided these reasoning chains are developed and expressed via *actions* upon developmentally appropriate *objects*.

Consider the following example of a fourth-grade student who developed a quite sophisticated variation of the "casting our nine's" strategy for checking division using computer graphic representations of the $Base_{10}$ blocks. This child had previously developed skill in using a paint program and had access to a set of clip art illustrating the standard $Base_{10}$ blocks[2] shown earlier.

To put this example into context, the child had just finished exploring a *teacher-posed problem* requiring creation of a method to determine whether any given number was divisible by 9. This task proved to be difficult, but solvable for the student. Encouraged by the earlier success and based on observations made during this effort, the child expanded the problem. After much additional work–much of it using a paint program on the classroom computer as an *active sketch* pad–the student claimed to be able to tell not only whether any given number was divisible by 9, but also what the value of any remainder might be. This expanded *self-posed problem* is the source of the following discussion.

Teacher: So you tell me you can predict the remainders when dividing by 9 for any number at all? This I have got to see.

Student: You see, all you have to do is just keep adding and adding and adding, till you can't add anymore. This final add-answer is the leftovers.

The term "leftovers" arose from earlier work where the $Base_{10}$ blocks were referred to as candies from a candy factory that had been packaged into boxes of various sizes. Although this problem had been posed in the realm of pure number, it is interesting to note how the child occasionally uses this earlier vocabulary. In this example, I have kept the child's language and not attempted to correct it.

Teacher: Let's see if you get this one right. Try 627 divided by 9.

Student: Watch! 6 and 2 is 8 and 7 is 15, but I can still add some more. One and 5 is 6, so the remainder is 6.

The student then proceeded to verify this result by performing the standard division procedure and showing the resulting remainder quite proudly (see Figure 9). Several further problems were given and the student was able to use the newly developed method to predict the remainder in each case, a prediction that was likewise verified.

Teacher: Well, I can see that you seem to have figured something out that works, but I have a problem with this. You seem to be just adding together numbers of different place values. See, you are

FIGURE 9. Student vertification using standard division.

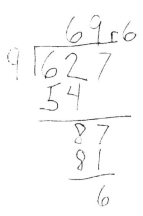

adding tens to ones and ones to hundreds, and that just does not seem right. Why can you do this? Also, how do you know your method works for all numbers? Maybe we just got lucky on these!

It must be mentioned that the students in this class were quite used to this type of extended question by this time and joked among themselves that, whenever their teacher questioned them in this manner, it meant that they were really on track of something cool. Given this new challenge to continue exploring this remainder prediction method, the student went back to the computer taking a set of $Base_{10}$ blocks along for reference purposes. The following came as a result of a great deal of student *actions* upon these two basic *objects* types.

Student: I think I can explain it to you now. My method will always work and I am not really adding ones to hundreds or anything like that. I'm just adding up the leftovers from the hundreds or the tens–or whatever.

Teacher: OK. Why don't you show me what you've figured out?

Student: We'll start with the ones. Whenever I look at a bunch of ones, it is easy to see the number of remainders–it's just the number we've got. See, because if we've got nine then it divides by nine; and, if we've got anything less than that, then that is the remainder.

Teacher: OK, but what about the tens and hundreds and all the rest?

Student: Look at the tens. I've drawn what it would look like here when I divide any ten by nine. See, I'll always have one one left. (See Figure 10.)

Student: And it does not stop there. Look, for each hundred I'm gonna have one one. (See Figure 11.)

Student: Because I first can divide up the hundred into nine tens with one ten left. I've already figured out what happens with the tens, but look. See, the ten can be divided up by nine and leaves one one.

Teacher: OK. But what about the thousands? After all, there are bigger numbers than these out there!

Student: Good try, but you can't fool me! The thousands is made up of ten hundreds. Each hundred will have one left over, which

FIGURE 10. Remainder for ten divided by nine.

FIGURE 11. Remainder for one hundred divided by nine.

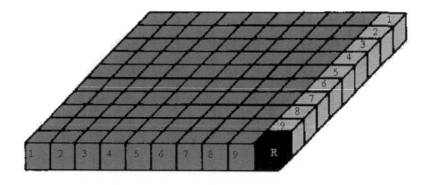

leaves one one each. So I can add up the ten ones leftover from each hundred, make a ten out of it, and I already know for a ten that I will have one one left. This will work all the way up the line! I will always have one one as a leftover no matter how big the number is.

As this example serves to illustrate, this student developed some very sophisticated chains of mathematically relevant reasoning based upon the *actions* taken upon developmentally appropriate *objects*. We can further see how the computer was used to enable performing actions on the *sketch* level object, such as dividing and labeling the respective shares, which would have been difficult to perform upon the *manipulative* itself. This example is also important since we see that the *manipulative*–far from being replaced with the *sketch*–served as an important referent for further work. This example captures not only the child's actions upon these two classes of objects, but also serves to document some of the actions typical of the transition between functioning at one level of object to another (Connell, 1995).

When we look more upon the nature and type of mathematical thinking and less upon the form in which that thinking is exhibited, I believe my statements hold up quite well. In particular, I have come to find that students are able to engage in mathematical thinking at a variety of developmental levels. Whether students are using *manipulatives*, computer-enhanced *sketches*, *mental pictures*, or formal *abstractions*, I have documented markedly similar patterns of thinking and of action upon each of these types of *objects* of thought.

As our example from casting out nines serves to illustrate, the sketch resulting from application of this model can come to have tremendous predictive power. Not only that, but often this predictive power is based upon the underlying physical manipulative. This is an extremely important point, for just as in the case of programming objects within an object-oriented programming environment, we observe that each of the developing *objects of thought* inherit much of their power and utility from the underlying, more primitive objects upon which they are based.

THE POWER OF THE COMPUTER SKETCH

The last point I would like to make is that, in adopting this *action upon objects* approach, technology has a natural place. I have suggested for some time that technology fits most naturally as a tool to enhance the development of the sketch and a mental picture level (Connell, 1998; Connell & Bounieav, 1997; Connell, 1995; Connell, 1994; Connell & Ravlin, 1988). These examples show how technology can be used to enhance the instructionally relevant properties of a physical sketch.

I have observed that technologically enhanced sketches, such as those described in these examples, share many of the characteristics normally associated with a *mental picture*. In particular, it is possible to act upon these in a very flexible, fluid fashion and one that is easily modified by the student. Unlike the mental picture, however, such technologically enhanced sketches have additional structure built into their very nature–having their foundation in a programmed computer object–so that the various deformations and mutations we perform upon these sketches do not alter their foundational nature.

It is in many ways a much better object than a traditional sketch to fall back upon in developing a powerful mental picture. For example, one cannot always keep constant control over the various parameters in a mental picture to the degree he or she can in using a computer object. Let's imagine that we have just created a square of size one by one (1 × 1) using any of the more common paint and sketch programs. Since this is, by definition, a unit square, it is easy to show that its area will be one square unit (see Figure 12).

Now, imagine that we mentally grab one of the corners and then stretch this unit square two additional units to the right so that its final dimensions are one by three (1 × 3; see Figure 13).

This resulting figure is a rectangle one unit on one side and three units on the other. This is extremely easy to do with a technologically enhanced sketch, and, in so doing, we can be assured that the relative proportions will always be exactly those that we desire. Furthermore, these relative proportions are maintained even when we enlarge or shrink the base image (see Figure 14).

Contrast this with a more traditionally formed mental image. It is easy to imagine that a student might be able to mentally imagine stretching a square but would not be able to maintain the exact proportions necessary to make accurate predictions regarding the resultant area.

FIGURE 12. Computer-based unit square.

FIGURE 13. Computer-based one-by-three rectangle.

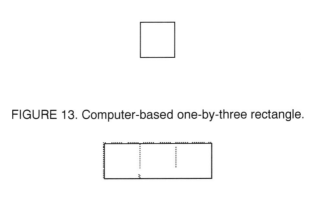

To make this point even more clearly, let us return to our original (1 × 1) unit square. Picture this square in your mind, and imagine grabbing it by one corner and stretching it so that it becomes a rectangle three units in one direction and five in the other (3 × 5). Although a student might be able to give you the correct response that the area of the resultant figure is 15 square units, it is highly unlikely that he or she would be able to mentally keep the correct proportions and ratios to the degree of accuracy that a sketch or paint program would be able to do (see Figure 15).

In many ways such computer-based objects, as an object of thought, are better suited than a traditional sketch or mental picture. It beats out a sketch due to its validity; it beats out a mental picture because of its rigidity.

FIGURE 14. Proportion invariance example.

FIGURE 15. Three-by-five computer sketch.

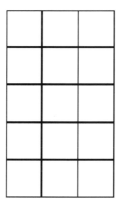

CONCLUSION

Since their inception in 1989, the NCTM *Standards for Curriculum and Evaluation* have been quietly (and in some cases quite vocally) changing the face of what constitutes mathematics and how we think about its teaching and learning. Consider the following: Mathematics is communication; mathematics is reasoning; mathematics is problem solving; and mathematics is connections. These statements are drawn directly from the process strands of the 1989 standards document and constitute a major revision as to what counts in mathematical thinking.

These process strands have become the de facto class of acceptable *actions* that are to be performed in mathematics education. These actions are much different in both form and substance than earlier procedural and content-driven *actions* of the past. And, it is in this very difference that we see that it is possible to perform these actions–to reason, problem solve, connect, and communicate mathematically–at every developmental age, provided that an appropriate *object* for this action to act upon is present.

With the addition of the *technology-enhanced object*, there is tremendous leveling of the playing field in terms of what counts as mathematical thinking. With this addition, we can observe that both the thinking processes and sophistication, as shown by students of various ages, begin to parallel each other. The technology serves an important role in this process. In particular, the computer can serve as a tool to record the information that has been generated by the students' activities. It can then capture the essence of the activity by allowing the students to organize their work in powerful structures. For example, a table might be built using developed data in a row-and-column structure, or an organizing graphic might be constructed reflecting the series of activities the children have performed. The technology can then be used to create formal records of action that may be shared or used in later problem-solving endeavors. These records of action can even be shared with others outside the immediate sociocultural milieu within which the student is working via the World Wide Web.

The computer, with its object-oriented interfaces and tools geared specifically to enable the user to perform specific actions upon specific objects, lends itself perfectly to an *action upon object* model of learning and instruction. Some possible models include step-by-step development of mental activities (Galperin & Talizyna, 1979; Leontiev, 1972), action reification (Sfard, 1994), and emergent structural theory (Connell, 1996), to name a few. This linkage between suggestions from learning theory and the object classes engendered by the technology is far too priceless to allow going to waste.

Of course, this is not the end. In imagining the years ahead and the development of new programming paradigms, we will no doubt discover more links between these increasingly powerful and plastic tools and our own cognitive processes. The form of object-orientation utilized in this writing is a relatively new construct, going back approximately 15 years. Who knows what new insights into human cognition we may find as we explore the next generation of technologies made in our own mental image?

In conclusion, the mathematics classroom of today is a far cry from that of 15 years ago. The technological objects upon which we act now routinely have intelligence built into them. If you made an error in using a slide rule, as was quite common when I was in school, all that would happen is that your result would be inaccurate. Today's intelligent objects have the potential of notifying you of the error, suggesting new options for you to consider, and quite possibly leading you astray through giving information at a level that does not match your understanding.

Ironically, the importance of understanding the underlying mathematical concepts in this scenario is significantly stronger than during the previous introduction of technology–the calculator. The calculator, despite the vast hue and cry of the time, proved to be a relatively benign invention, helping, as it were, with the numeric processing of skills and procedures that the child would have to construct, apply, and evaluate. The computer, with its increasingly powerful objects of thought, is a much more insidious problem.

On one hand, it allows us to leverage our thinking tremendously forward through interaction with tools which themselves have rudimentary problem-solving abilities. Furthermore, the natures of the data organization made possible through using technology lend themselves to types of approaches with known mathematical pay-off. On the other hand, however, we are in a very real danger of having our tools become more intelligent than the people using them. It is the responsibility of mathematics educators everywhere to ensure that our teacher candidates, teachers, and students are able to *use* the *tools* and not be *used* by *them*.

NOTES

1. I am using the term *primitive* to refer to a foundational material from which later concepts will be built–not necessarily as referring to simple or unsophisticated. This is in keeping with the use of *primitive* in computer systems. Thus, Dienes Base$_{10}$ blocks might well be considered as an experiential primitive upon which later conceptualizations would be built.

2. As an aside, the clip art used by the student was the same as that which I have used in the writing of this chapter. This pretty well guarantees the fidelity of the included graphics to say the least.

REFERENCES

Adams, D. (1983). *Life, the universe, and everything*. New York: Pocket Books.

Bouniaev, M., & Connell, M. L. (1999). Constructivism and SSDMA as a basis for technology use in mathematics teacher education. In J. D. Price, J. Willis, D. Willis, M. Jost, & S. Boger-Mehall (Eds.), *Technology and teacher education annual 1999* (pp. 945-950). Charlottesville, VA: Association for the Advancement of Computing in Education.

Brownell, G., & Brownell, N. (1998). Building bridges: Technology and curriculum theory and practice. In S. McNeal, J. D. Price, S. Boger-Mehall, B. Robin, & J. Willis (Eds.), *Technology and teacher education annual 1998* (pp. 396-399). Charlottesville, VA: Association for the Advancement of Computing in Education.

Connell, M. L. (1998). Technology in constructivist mathematics classrooms. *Journal of Computers in Mathematics and Science Teaching, 17*(4), 311-338.

Connell, M. L. (1996). *Consultancy report of IKIP Malang*. Report prepared for and presented to the Indonesian Ministry of Higher Education. (Available from Michael L. Connell, PhD, 344 FAH, University of Houston, Houston, TX).

Connell, M. L. (1995). A constructivist use of technology in pre-algebra. In Douglas Owens (Ed.), *Psychology of mathematics education* (pp. 187-192). Columbus, OH: Ohio State University.

Connell, M. L. (September, 1994). *Making mathematics: Teaching and learning mathematics in a technologically rich environment*. Paper presented to the 75th anniversary research symposium of the University of Illinois, Champaign-Urbana, IL.

Connell, M. L. (1988). Using microcomputers in providing referents for elementary mathematics. In M. Miller-Gerson (Ed.), *The emerging frontier: Interactive video, artificial intelligence and classroom technology* (pp. 55-60). Phoenix, AZ: Arizona State University.

Connell, M. L. (1986). *Conceptual based evaluation and student tracking in elementary mathematics*. Unpublished master's dissertation, pp. 75-80. University of Utah, Salt Lake City, UT.

Connell, M. L., & Abramovich, S. (1999). New tools for new thoughts: Effects of changing the "tools-to-think-with" on the elementary mathematics methods course. In J. D. Price, J. Willis, D. Willis, M. Jost, & S. Boger-Mehall (Eds.), *Technology and teacher education annual 1999* (pp. 1052-1058). Charlottesville, VA: Association for the Advancement of Computing in Education.

Connell, M. L., & Bounieav, M. (1997). Shared recommendations of SSDMA and constructivism upon technology in mathematics education. *The Researcher 12*(1), 21-28.

Connell, M. L., & Ravlin, S. B. (April, 1988). *The flagpole factory: Providing a referent for linear equations*. Paper presented at the annual meeting of the International Association of Computers in Education, New Orleans, LA.

Du Plessis, J. P. (1995). A model for intelligent computer-aided education systems. *Computers & Education. 24*(2), 89-106.

Fagan, P. J., & Thompson, A. D. (1989). Using a database to aid in learning the meanings and purposes of mathematical notations and symbols. *Journal of Computers in Mathematics and Science Teaching, 8*(4), 26-30.

Galperin, P., & Talizyna T. (1979). Contemporary condition of the SSDMA theory. *Vestnik, MGU, Psychology, 14*(4), 54-63.

Garofalo, J., Shockey, T., & Drier, H. (1998). Guidelines for developing mathematical activities incorporating technology. In S. McNeal, J. D. Price, S. Boger-Mehall, B. Robin, & J. Willis (Eds.), *Technology and teacher education annual 1998* (pp. 396-399). Charlottesville, VA: Association for the Advancement of Computing in Education.

Harvey, F. A., & Charnitski, C. W. (1998, February). *Improving mathematics instruction using technology: A Vygotskian perspective.* (ERIC Document Reproduction Service No. 423 837)

Leontiev, A. N. (1972). *Problems of psychological development.* Moscow, USSR: Moscow State University Publishing House.

National Council of Teachers of Mathematics, Commission on Standards for School Mathematics. (1989). *Curriculum and evaluation standards for school mathematics.* Reston, VA: Author.

Peck, D. M., & Connell, M. L. (1991). Using physical materials to develop mathematical intuition. *Focus on Learning Issues in Mathematics, 13*(4), 3-12.

Sfard, A. (1994). Reification as the birth of metaphor. *For the Learning of Mathematics, 14*(1), 44-55.

Sfard, A. (1991). On the dual nature of mathematical conceptions: Reflections on processes and objects as different sides of the same coin. *Educational Studies in Mathematics, 22*(1), 1-36.

Whitehead, A. N. (1978). Process and reality: An essay in cosmology. In D. R. Griffin & D. W. Sherburne (Eds.), *Process and reality: Corrected edition* (p. 267). New York: The Free Press. (Original work published in 1929).

Sharon Dugdale

Pre-Service Teachers' Use
of Computer Simulation
to Explore Probability

SUMMARY. Knowledge of probability is fundamental to making well-reasoned decisions, and there is a need for more attention to both students' learning of probability and teachers' preparation in this area. This paper examines pre-service K-6 teachers' reasoning and the role of a computer in their investigation of a game of chance. Proceeding from experiments to theoretical probability, participants engaged in a combination of small group work, computer simulation, and whole-class discussion. Participants' reasoning suggests that the computer's easy generation of repeated long runs of events might foster insights into probability that are different from those likely to emerge from more limited experimental trials. *[Article copies available for a fee from The Haworth Document Delivery Service: 1-800-342-9678. E-mail address: <getinfo@haworthpressinc.com> Website: <http://www.HaworthPress.com> © 2001 by The Haworth Press, Inc. All rights reserved.]*

KEYWORDS. Probability, computer simulation, pre-service teachers, mathematical reasoning

Whether we are trying to understand insurance plans and medical testing, making decisions about playing the lottery, or simply interpreting the weather

SHARON DUGDALE is Professor, Division of Education, University of California, One Shields Avenue, Davis, CA 95616 (E-mail: ssdugdale@ucdavis. edu).

[Haworth co-indexing entry note]: "Pre-Service Teachers' Use of Computer Simulation to Explore Probability." Dugdale, Sharon. Co-published simultaneously in *Computers in the Schools* (The Haworth Press, Inc.) Vol. 17, No. 1/2, 2001, pp. 173-182; and: *Using Information Technology in Mathematics Education* (ed: D. James Tooke and Norma Henderson) The Haworth Press, Inc., 2001, pp. 173-182. Single or multiple copies of this article are available for a fee from The Haworth Document Delivery Service [1-800-342-9678, 9:00 a.m. - 5:00 p.m. (EST). E-mail address: getinfo@haworthpressinc.com].

173

forecast, our knowledge of probability affects our everyday lives. Ability to deal with probability is important in making well-reasoned decisions as citizens and as consumers. In short, "There is perhaps no other branch of the mathematical sciences that is as important for *all* students, college bound or not, as probability and statistics" (Shaughnessy, 1992, p. 466).

In recognition of the role of probability in daily life, as well as in many careers, the *Curriculum and Evaluation Standards for School Mathematics* (National Council of Teachers of Mathematics, 1989) advocates probability as an important strand of the mathematics curriculum from the primary grades through high school. The more recent *Principles and Standards for School Mathematics* (NCTM, 2000) further emphasizes the need for all students to understand and apply basic notions of chance and probability. In particular, students in grades three through five should learn about probability as a measurement of the likelihood of events, and students in grades six through eight should encounter numerous opportunities to engage in probabilistic thinking and develop notions of chance. These experiences build the foundation for more formal study of probability in grades nine through twelve.

Bright and Harvey (1981) suggest that judging the fairness of games is a good first step in understanding basic concepts of probability. In an experimental study with 82 seventh graders, these researchers found that students' ability to determine the fairness of a game improved significantly through experience playing games of chance before engaging in any formal instruction. In a similar vein, Shaughnessy and Bergman (1993) recommend that the teaching of probability be activity-based and proceed from experiments to theory. Further, they argue that simulations are an especially valuable problem-solving tool in this context, and that understanding of probability is best developed through small group work and frequent discussions.

As with other areas of mathematics, computers offer new ways of approaching probability in the classroom. For example, after students obtain sufficient experience gathering data with physical objects, they can then use a computer simulation to streamline the process by quickly producing large runs of experimental trials, such as coin tosses or cube rolls. Further, the software can offer various representations of the data, tally specific outcomes, and perform other analysis tasks.

Successful implementation of probability learning experiences in classrooms relies on teachers' understanding and preparation. Only recently have curriculum development projects made inroads into the teaching of probability and statistics in the middle grades, and very few students encounter substantial instruction on these topics in high school (Shaughnessy, 1992). Moreover, a large proportion of students do not understand many of the basic statistical concepts they have studied (Garfield & Ahlgren, 1988). Shaughnessy and Bergman (1993) point out that the experience in this area for most of our classroom teachers is minimal. Further, they contend that "whatever probability and statistics that these teachers have acquired usually was not taught in a way to develop understanding (simulations, activity-based, frequent discussions)" (p. 193). To improve students' experiences in learning about probability, these researchers see a need to better understand teachers' conceptions of probability and the roles that computer software can play in supporting the learning of probability.

This paper discusses pre-service K-6 teachers' approaches to a probability problem and the role of the computer in their solution process. Given the task of modifying number cubes to make a game of chance fair, the pre-service teachers used computer simulation to test their games and work on questions of fairness.

PRELIMINARY EXPERIENCES

Following some basic coin tossing activities and discussion of randomness, a class of sixteen K-6 pre-service teachers were introduced to computer software (Edwards, 1991) that simulated probability experiments with coins, cubes, and spinners. Some introductory activities with the software familiarized the participants with the options for setting up experiments and viewing outcome tallies, charts of relative frequencies, bar graphs, and line graphs. Long runs of simulated coin tosses were compared with data gathered from the activities already carried out with real coins, and the class discussed the role of the computer as a convenient way to simulate gathering large amounts of data quickly.

Class members used the software to perform some simple experiments with coins and cubes. The software allowed the selection of particular events to investigate. For example, in rolling two cubes, it was possible to focus on events such as rolling doubles, having the sum of the cubes be even, or having the sum be a particular number. The software tallied the requested results and computed relative frequencies.

PROBLEM SETUP

When participants were comfortable with the features of the software, the instructor proposed a game: A pair of cubes would be rolled 200 times; the class would get a point each time the sum of the cubes was even, and the instructor would get a point each time the sum was odd. The simulation was run, and the instructor won by a few points. Repeating the experiment several times resulted in some wins for the class and some for the instructor. When asked if they would prefer to receive a point when the sum was odd instead of even, the participants decided that it didn't matter. This led to some discussion of what it means for a game to be "fair."

The instructor proposed playing the game again, only this time with points awarded based on whether the *product* of the two cubes was even or odd. Given the choice of either the even or the odd product outcome, participants were divided, with some votes for each choice and many who preferred to wait and see the results before committing to which outcome they preferred. When running the simulation produced results dramatically in favor of the even product, participants were quick to pronounce the game unfair.

The software allowed the creation of a "custom cube" (i.e., a cube with a non-standard combination of numbers on the six sides). So, for example, a cube could be made to have two faces numbered 4 and no face numbered 5. The instructor challenged the class to design a cube so that "product even or odd" would be a fair game. Working in groups of four, with one computer per group, class members investigated the problem and recorded their results on overhead transparencies in preparation for sharing with the class. In rolling two cubes, a constraint of the software was that the two cubes be identical. Participants were invited to address the same problem with non-identical cubes after resolving it for identical cubes.

SOLUTION PROCESS

As noted above, the relative frequencies of even and odd products for a pair of normal cubes suggested a substantial advantage for the player who won a point when the product was even. This observation influenced all groups to begin their efforts by changing some of the even-numbered faces on a normal cube to odd numbers, in order to have more odd faces and therefore more odd products. The strategy used by each group was to construct a cube, try a large run of simulated rolls, and look at the relative frequencies. If the first run showed somewhat close relative frequencies, participants tended to repeat the trial to determine whether the results were consistently close. All groups soon concluded that the game was most fair with cubes that had two even faces and four odd faces. This combination produced the most nearly equal relative frequencies. Each group then focused on deciding whether the game with these cubes was truly fair.

From this point, the groups diverged in their thinking. One participant, Jan, set the direction of her group's inquiry by noting that the relative frequencies were quite close, and even though it wasn't certain that the game was completely fair, it was "fair enough." When asked, "If you played the game with someone, would you care which outcome you got–whether you got a point for an even product or an odd product?" she responded, "No. I would be willing to take either even or odd, whatever the other person didn't want. It's close enough for me."

Jan's group ran several more trials, noting that the relative frequencies were consistently close, and sometimes very close. There was some speculation that the differences in the relative frequencies from one run to another might be typical of the variation that could be expected in a fair game. This line of reasoning seemed about to prevail until another member of the group, Chris, noted that the repeatedly close trials uniformly favored the same outcome: After each long run of rolls, the outcome with the higher frequency was always the even product. Chris argued that a fair game shouldn't have the same winner every time, even if the scores are always close.

Performing several further runs, the group had to agree that this was problematic, and it challenged the notion that being close made it "fair enough." This line of reasoning also brought to light a contrast between the result of a single roll and the result of a tally of a large number of rolls. If the "game" was to roll the cube once and see who wins, then Jan was still willing to take either outcome, even though her opponent might get a slight advantage. However, if the game was to tally 1,000 rolls and declare the winner to be the outcome with the highest frequency, then Jan was no longer willing to take either outcome. She suggested that, if the game prize, whatever it was, say a bag of jelly beans, could be divided by giving the winner one jelly bean on each roll of the cubes, then at the end of a long run of rolls, the player who won when the product was even would have a few more jelly beans, but not enough to bother her if she were the other player. However, if they waited until the end of a long run of rolls and awarded the whole bag of jelly beans to the player with the most wins, then the player who chose the odd product wouldn't get anything.

In sharing their results with the class (see Figure 1), this group presented their cube with two even faces and four odd faces as the best combination, and concluded that the game was not entirely fair because, with large numbers of rolls, it was clear that even products occurred more often.

Another group, faced with the quandary of whether their cube with two even faces and four odd faces made the game fair, resolved the issue by constructing a matrix of all possible outcomes and verifying that there were, indeed, more even product outcomes than odd product outcomes. As their work shows in Figure 2, of 36 equally likely outcomes, 20 were even and 16 were odd.

This group's matrix provided a framework in which they could conveniently check theoretical probabilities for other cases (e.g., cubes with one even face and five odd faces). In presenting their work (Figure 2) to the class, one of the group members, Stacy, explained, "This cube has too many even outcomes, but if we change one of the even faces to odd, it [the matrix] would

FIGURE 1. Conclusions shared by Jan's group.

Fairest Cube

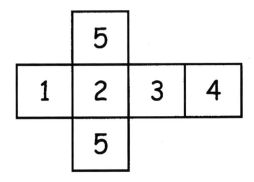

Relative Frequency For "even"
.569, .523, .552, .546, .563, .528

One roll: not quite fair

Tally of 1000 rolls: <u>REALLY</u> unfair!
"even" <u>ALWAYS</u> wins!

end up with only one row and one column of even outcomes, so then there would be 11 even outcomes to 25 odd outcomes, and that would be even less fair, so we stayed with the cube with two even faces and four odd faces." Interestingly, this was the only one of the four groups to arrive at a systematic representation of the possible outcomes or to produce a theoretical probability.

Some of the groups tackled the question of making the game fair with non-identical cubes. Those who had noted an apparent bias toward even products using cubes with two even faces and four odd faces thought of altering one of the cubes to have one even face and five odd faces. They reasoned that this would produce more odd products, but not as many odd products as having both cubes with one even face and five odd faces. Stacy's group, whose work is shown in Figure 2, used a similar outcome matrix to assess the fairness of this combination of two non-identical cubes and found

FIGURE 2. Conclusions shared by Stacy's group.

Cube with 2 even faces and 2 odd faces

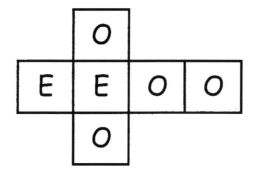

Possible Outcomes

	E	E	O	O	O	O
E	E	E	E	E	E	E
E	E	E	E	E	E	E
O	E	E	O	O	O	O
O	E	E	O	O	O	O
O	E	E	O	O	O	O
O	E	E	O	O	O	O

20 even : 16 odd

Not Fair

that it favored odd products 20:16. At this point they began to focus on what they wanted to appear in the outcome matrix. They decided that a matrix with three rows even and three rows odd would be a simple way of getting an equal number of even and odd outcomes, and this led them to propose one cube with all odd faces and the other cube half even and half odd.

Other groups proceeded less systematically. The software had no option for creating two non-identical cubes, so when participants began considering non-identical cubes, the software could no longer be used to test long runs of rolls. Hence, groups who were still depending on the software to assess fairness had difficulty proceeding with their non-identical cubes. Lacking the facility for testing long runs of rolls with non-identical cubes, one of these groups abandoned their ideas about selectively modifying the cubes they had settled on as most fair if the cubes had to be identical. After some discussion of possibilities, they proposed using one normal cube with faces numbered 1 through 6 and one cube with all faces numbered 1. Their argument was simple: The normal cube had three odd-numbered faces and three even-numbered faces, so rolling it alone would give an equal chance of getting an odd or an even number. With the second cube always rolling a 1, the product of the two cubes would equal the roll of the first cube. Hence, the game of "product even or odd" would be a fair game with these cubes.

DISCUSSION

The solution process summarized above illustrates a range of sophistication in pre-service K-6 teachers' understanding of probability and their approaches to a problem of chance. With the aid of the computer's simulated cube rolls and relative frequencies, all four groups of participants succeeded in identifying the fairest possible cube for the game of "product even or odd." The two groups whose work was discussed above were the most successful in constructing convincing arguments that the game was not fair with this cube, and only one group produced a systematic list of outcomes and a theoretical probability (Figure 2). However, all groups questioned the fairness of the game and perceived a need for a more logical argument than could be deduced from the cube rolling and data summarizing facility of the software.

Although the computer's rapid data generation capabilities made it possible to run thousands of simulated trials, tally the outcomes, and compute the relative frequencies, the computer results alone did not bring the problem to a satisfying closure. When it was time to share results, only Stacy's group appeared confident that they thoroughly understood the problem situation. Even though Jan's group felt sure that the game resulted in more even products than odd for long runs of rolls, they were unsure about why this happened, and they had in common with the other two groups a sense that they had not quite achieved a complete conclusion.

The concluding activity of sharing and discussing solutions from each of the groups brought together a combination of ideas that enriched the learning experience for all participants. Groups that had not succeeded in bringing a clear closure to their investigation were able to see ways to do that. They agreed that the systematic listing of outcomes and production of a theoretical probability by Stacy's group afforded an advantage in understanding, and even predicting, the results of rolling cubes with any combination of faces. Participants noted that Stacy's group had moved easily from testing two identical cubes to considering a pair of non-identical cubes; whereas, for the groups who were relying heavily on the computer's data generation, the problem extension to a case that the software could not handle increased their sense that they needed to understand the situation in a more fundamental way than seeing experimental trials.

Even with the substantial contribution of Stacy's group to the class discussion, the learning experience was not one-sided. The presentation of results from Jan's group introduced aspects of the situation that had not occurred to Stacy's group (or to any other group). By the end of the solution sharing and discussion, the class as a whole had developed a broader perspective of the problem and had distinguished what information can be attained from experimental trials and from theoretical probabilities.

Given the apparent consensus that it was desirable to find a means of producing a theoretical probability, one might reasonably ask whether the extensive computer-generated data collection added value beyond what could be done easily by rolling real cubes. The outcome matrix and theoretical probability (Figure 2) produced by Stacy's group are typical of what teachers hope will result from this sort of probability activity, and this solution could arguably have come about nearly as easily without the computer. However, the line of reasoning and resulting insight experienced by Jan's group undoubtedly depended on the computer's facility for producing repeated long runs. It was the consistent results of each of a large number of long runs that caught Chris's eye and challenged Jan's acceptance of the game as "fair enough." Chris's observation rescued the group from concluding that the game was fair on the basis that the relative frequencies were close.

CONCLUSION

This paper has examined the reasoning of pre-service K-6 teachers as they grappled with a problem of chance. As often happens in small group work, the interactions and negotiations within the groups served to challenge misconceptions and hone participants' reasoning. The subsequent sharing and discussion with the class as a whole further developed participants' perceptions of the problem situation they faced; and, in the end, the class achieved a clearer sense of what it means for a game to be fair, what can be learned from a large number of experimental trials, and the usefulness of a theoretical probability.

The computer played an important role in generating simulated data, tallying outcomes, and computing relative frequencies. However, participants did not find this information sufficient to bring the problem to closure. A systematic listing of outcomes and a theoretical probability served that purpose. Beyond the immediately apparent contributions of the computer to this type of investigation, the reasoning engaged in by at least one group suggests that the computer's easy generation of repeated long runs of events might foster insights into probability that are different from those likely to emerge from more limited experimental trials.

REFERENCES

Bright, G. W., & Harvey, J. G. (1981). Fair games, unfair games. In A. P. Shulte & J. R. Smart (Eds.), *Teaching statistics and probability* (1981 Yearbook, pp. 49-59). Reston, VA: National Council of Teachers of Mathematics.

Edwards, L. (1991). *A chance look* [Computer software]. Pleasantville, NY: Sunburst.

Garfield, J., & Ahlgren, A. (1988). Difficulties in learning basic concepts in probability and statistics: Implications for research. *Journal for Research in Mathematics Education, 19*(1), 44-63.

National Council of Teachers of Mathematics. (1989). *Curriculum and Evaluation Standards for School Mathematics.* Reston, VA: Author.

National Council of Teachers of Mathematics. (2000). *Principles and Standards for School Mathematics.* Reston, VA: Author.

Shaughnessy, J. M. (1992). Research in probability and statistics: Reflections and directions. In D. A. Grouws (Ed.), *Handbook of research on mathematics teaching and learning* (pp. 465-494). New York: Macmillan.

Shaughnessy, J. M., & Bergman, B. (1993). Thinking about uncertainty: Probability and statistics. In P. S. Wilson (Ed.), *Research ideas for the classroom: High school mathematics* (pp. 177-197). New York: Macmillan.

Index

Printed and bound by CPI Group (UK) Ltd, Croydon, CR0 4YY

17/10/2024

01775685-0014